MW01060406

Understandable Statistics
Concepts and Methods

ELEVENTH EDITION

Charles Henry Brase
Regis University-Denver

Corrinne Pellillo Brase
Arapahoe Community College

Prepared by

Joseph Kupresanin
Cecil College

CENGAGE
Learning

Australia • Brazil • Mexico • Singapore • United Kingdom • United States

For product information and technology assistance, contact us at **Cengage Learning Customer & Sales Support, 1-800-354-9706**.

For permission to use material from this text or product, submit all requests online at **www.cengage.com/permissions**
Further permissions questions can be emailed to **permissionrequest@cengage.com**.

ISBN-13: 978-1-285-46283-7
ISBN-10: 1-285-46283-1

Cengage Learning
200 First Stamford Place, 4th Floor
Stamford, CT 06902
USA

Cengage Learning is a leading provider of customized learning solutions with office locations around the globe, including Singapore, the United Kingdom, Australia, Mexico, Brazil, and Japan. Locate your local office at: **www.cengage.com/global**.

Cengage Learning products are represented in Canada by Nelson Education, Ltd.

To learn more about Cengage Learning Solutions, visit **www.cengage.com**.

Purchase any of our products at your local college store or at our preferred online store **www.cengagebrain.com**.

Printed in the United States of America
1 2 3 4 5 6 7 17 16 15 14 13

Table of Contents

Chapter 1: Getting Started

Section 1.1

1. Individuals are people or objects included in the study, while a variable is a characteristic of the individual that is measured or observed.

3. A parameter is a numerical measure that describes a population. A statistic is a numerical value that describes a sample.

5. **(a)** These numerical assignments are at the nominal level. There is no apparent ordering in the responses.
   (b) These numerical assignments are at the ordinal level. There is an increasing relationship from worst to best levels of service. These assignments are not at the interval or ratio level. The distances between numerical responses are not meaningful. The ratios are also not meaningful.

7. **(a)** Meal ordered at fast-food restaurants.
   (b) Qualitative
   (c) All U.S. adult fast-food consumers.

9. **(a)** Nitrogen concentration (milligrams of nitrogen per liter of water).
   (b) Quantitative.
   (c) The water in the entire lake.

11. **(a)** Ratio. **(b)** Interval. **(c)** Nominal. **(d)** Ordinal. **(e)** Ratio. **(f)** Ratio.

13. **(a)** Nominal. **(b)** Ratio. **(c)** Interval. **(d)** Ordinal. **(e)** Ratio. **(f)** Interval.

15. **(a)** Answers vary. Ideally, weigh the packs in pounds using a digital scale that has tenths of pounds for accuracy.
   (b) Some students may refuse to have their backpacks weighed.
   (c) Informing students before class may cause students to remove items before class.

Section 1.2

1. In stratified samples, we select a random sample from each stratum. In cluster sampling, we randomly select clusters to be included, and then each member of the cluster is sampled.

3. Sampling error is the difference between the value of the population parameter and the value of the sample statistic that stems from the random selection process. The term is being used incorrectly here. Certainly larger boxes of cereal will cost more than smaller boxes of cereal.

5. No. Even though the sample is random, some students younger than 18 or older than 20 may not have been included in the sample.

7. **(a)** Stratified.
   (b) No. Each pooled sample would have 100 season ticket holders for men's basketball games and 100 season ticket holders for women's games. Samples with, for example, 125 and 75 tickets holders, respectively, are not possible.
   (c) Assign numbers 1, 2, …, 40 to the students and use a random-digits table or a computer package to draw random numbers.

9. Simply use a random digits table or a computer package to randomly select four students from the class.
   (a) Answers vary. Perhaps they are excellent students who make an effort to get to class early.

    (b)   Answers vary. Perhaps they are busy students who are never on time to class.

    (c)   Answers vary. Perhaps students in the back row are introverted.

    (d)   Answers vary. Perhaps taller students are healthier.

11. Answers vary.   **13.** Answers vary.

15. **(a)**   Yes, it is appropriate, as a number can repeat itself once it has occurred. The outcome on the fourth roll is 2.

    (b)   We will most certainly not get the same sequence of outcomes. The process is random.

17. Answers vary. Use single digits on the table to determine the placement of correct answers.

19. **(a)**   Simple random sampling. Every sample of size n from the population has an equal chance of being selected, and every member of the population has an equal chance of being included in the sample.

    (b)   Cluster sampling. The state, Hawaii, is divided into ZIP Codes. Then, within each of the 10 selected ZIP Codes, all businesses are surveyed.

    (c)   Convenience sampling. This technique uses results or data that are conveniently and readily obtained.

    (d)   Systematic sampling. Every fiftieth business is included in the sample.

    (e)   Stratified sampling. The population was divided into strata based on business type. Then a simple random sample was drawn from each stratum.

Section 1.3

1. Answers vary. People with higher incomes will likely have high-speed Internet access, which will lead to spending more time online. Spending more time online might lead to spending less time watching TV. Thus, spending less time watching TV cannot be attributed solely to high income or high-speed internet access.

3. No, the respondents do not constitute a random sample from the community for a number of reasons. For instance, the sample frame includes only those at the farmer's market, and Jill may not have approached people with large dogs or those who were busy, and participation was voluntary. Jill's T-shirt may have influenced responses.

5. **(a)**   No, those aged $18-29$ in 2006 became aged $20-31$ in 2008. The study is looking at the same generation.

    (b)   1977 to 1988, inclusive.

7. **(a)**   This is an observational study. The data collection method did not influence the outcome.

    (b)   This is an experiment. A treatment was imposed on the sheep in order to prevent heartworm.

    (c)   This is an experiment. The restrictions on fishing possibly led to a change in the length of trout in the river.

    (d)   This is an observational study. The data was collected without influencing the turtles.

9. **(a)**   Use randomization to select ten calves to inoculate with the vaccine. After a period of time, test all calves for the infection. No placebo is being used.

    (b)   Use randomization to select nine schools to visit. After ten weeks, survey students in all 18 schools for their views on police officers. No placebo is being used.

    (c)   Use randomization to select 40 subjects to use the skin patch. A placebo is used for the other 35 subjects. At the end of the trial, survey all 75 subjects about their smoking habits.

11. Based on the information, scheme A will be better because the blocks are similar. The plots bordering the river should be similar, and the plots away from the river should be similar.

Chapter Review Problems

1. Using a random-number table to select numbers for a Sudoku puzzle would be very inefficient. It would be much better to look at existing numbers that meet the puzzle's requirements and eliminate numbers that don't work.

3. **(a)** Stratified.
    (b) Students on your campus with work-study jobs.
    (c) Number of hours scheduled to work each week; Quantitative; Ratio.
    (d) Applicability to future employment goals, as measured by the scale given; Qualitative; Ordinal.
    (e) Statistic.
    (f) The nonresponse rate is 60%, and yes, this could introduce bias into the results. Answers vary.
    (g) No, since the students were only drawn from one campus, then the results of the study would only generalize to that campus, if the data were collected using randomization.

5. Using the random-number table, pick seven digits at random. Digits 0, 1, and 2 can correspond to "Yes," and digits 3, 4, 5, 6, 7, 8, and 9 can correspond to "No." This will effectively simulate a random draw from a population with 30% TIVO owners.

7. **(a)** This was an observational study because the researchers did not apply a treatment.
    (b) This was an experiment because the two groups were given different tests and the results were compared.

9. Answers vary. Questions should be worded in a clear, concise, and unbiased manner. No questions should be misleading. Commonsense rules should be stated for any numerical answers.

11. **(a)** This is an experiment; the treatment was the amount of light given to the colonies.
    (b) The control group is the colony exposed to normal light, while the treatment group is the exposed to continuous light.
    (c) The number of fireflies living at the end of 72 hours.
    (d) Ratio.

Chapter 2: Organizing Data

Section 2.1

1. Class limits are possible data values, and they specify the span of data values that fall within a class. Class boundaries are not possible data values; they are values halfway between the upper class limit of one class and the lower class limit of the next class.

3. The classes overlap. A data value such as 20 falls into two classes.

5. Width $= \dfrac{82-20}{7} \approx 8.86$, so round up to 9. The class limits are $20-28, 29-37, 38-46, 47-55, 56-64,$ $65-73, 74-82.$

7. (a) The distribution is most likely skewed right, with many short times and only a few long wait times.
    (b) A bimodal distribution might exist if there are different wait times during busy versus slow periods. During the morning rush, many long wait times might occur, but during the slow afternoon, most wait times will be very short.

9. (a) Yes.

    (b)
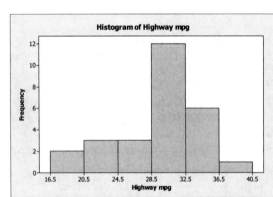

11. (a) The range of data seem to fall from 7 to 13 with the bulk of the data between 8 and 12.
    (b) All three histograms are somewhat mound-shaped with the top of the mound between 9.5 and 10.5 In all three histograms, the bulk of the data fall between 8 and 12.

13. (a) Because there are 50 data values, divide each cumulative frequency by 50 and convert to a percent.
    (b) 35
    (c) 6
    (d) 2%

15. (a) Class width = 25

(b)

Class Limits	Class Boundaries	Midpoints	Frequency	Relative Frequency	Cumulative Frequency
236–260	235.5–260.5	248	4	0.07	4
261–285	260.5–285.5	273	9	0.16	13
286–310	285.5–310.5	298	25	0.44	38
311–335	310.5–335.5	323	16	0.28	54
336–360	335.5–360.5	348	3	0.05	57

(c)

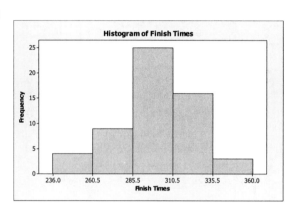

(d)

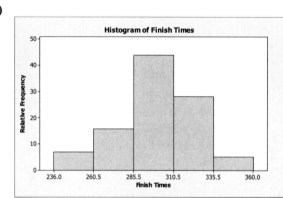

(e) This distribution is slightly skewed to the left but fairly mound-shaped, symmetric.

(f)

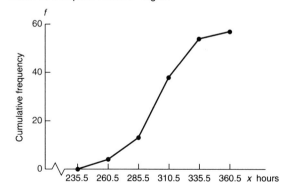

Chapter 2: Organizing Data

17. (a) Class width = 12

(b)

Class Limits	Class Boundaries	Midpoint	Frequency	Relative Frequency	Cumulative Frequency
1–12	0.5–12.5	6.5	6	0.14	6
13–24	12.5–24.5	18.5	10	0.24	16
25–36	24.5–36.5	30.5	5	0.12	21
37–48	36.5–48.5	42.5	13	0.31	34
49–60	48.5–60.5	54.5	8	0.19	42

(c)

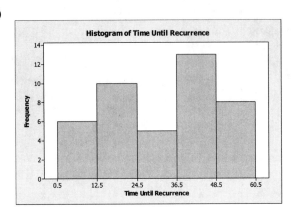

(d)

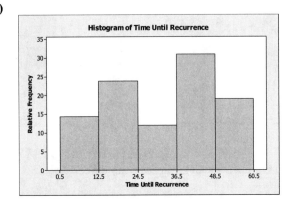

(e) The distribution is bimodal.

(f) To create the ogive, place a dot on the *x* axis at the lower class boundary of the first class, and then, for each class, place a dot above the upper class boundary value at the height of the cumulative frequency for the class. Connect the dots with line segments.

19. (a) Class width = 9

(b)

Class Limits	Class Boundaries	Midpoint	Frequency	Relative Frequency	Cumulative Frequency
10–18	9.5–18.5	14	6	0.11	6
19–27	18.5–27.5	23	26	0.47	32
28–36	27.5–36.5	32	20	0.36	52
37–45	36.5–45.5	41	1	0.02	53
46–54	45.5–54.5	50	2	0.04	55

© 2015 Cengage Learning. All Rights Reserved. May not be scanned, copied or duplicated, or posted to a publicly accessible website, in whole or in part.

(c)

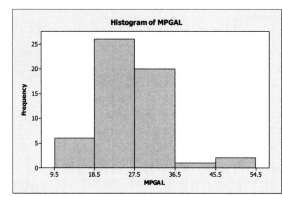

(d)

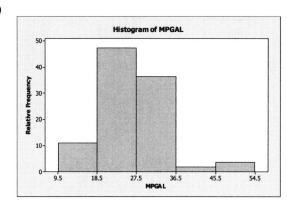

(e) This distribution is skewed right.

(f) To create the ogive, place a dot on the x axis at the lower class boundary of the first class, and then, for each class, place a dot above the upper class boundary value at the height of the cumulative frequency for the class. Connect the dots with line segments.

21. **(a)** Multiply each value by 100.

(b)

Class Limits	Class Boundaries	Midpoint	Frequency
46–85	45.5–85.5	65.5	4
86–125	85.5–125.5	105.5	5
126–165	125.5–165.5	145.5	10
166–205	165.5–205.5	185.5	5
206–245	205.5–245.5	225.5	5
246–285	245.5–285.5	265.5	3

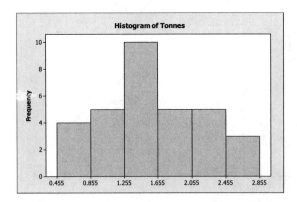

(c)

Class Limits	Class Boundaries	Midpoint	Frequency
0.46–0.85	0.455–0.855	0. 655	4
0.86–1.25	0.855–1.255	1.055	5
1.26–1.65	1.255–1.655	1.455	10
1.66–2.05	1.655–2.055	1.855	5
2.06–2.45	2.055–2.455	2.255	5
2.46–2.85	2.455–2.855	2.655	3

23. **(a)** 1
    (b) About $5/51 = 0.098 = 9.8\%$
    (c) 650 to 750

25.

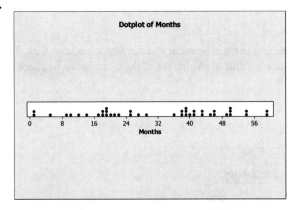

The dotplot shows some of the characteristics of the histogram, such as the concentration of most of the data in two peaks, one from 13 to 24 and another from 37 to 48. However, the dotplot and histogram are somewhat difficult to compare because the dotplot can be thought of as a histogram with one value, the class mark (i.e., the data value), per class. Because the definitions of the classes (and therefore the class widths) differ, it is difficult to compare the two figures.

Section 2.2

1. **(a)** Yes, since the percentages total more than 100%.
   (b) No. In a circle graph, the percentages must total 100%.
   (c) Yes. The graph is organized from most frequently selected to least frequently selected.

3. A Pareto chart because it shows the five conditions in their order of importance to employees.

5.

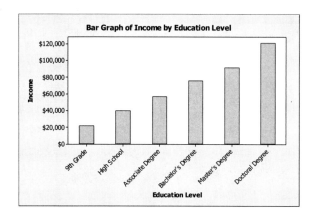

7.

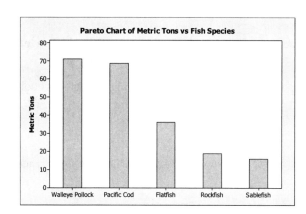

9.

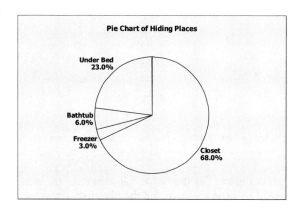

11. (a)

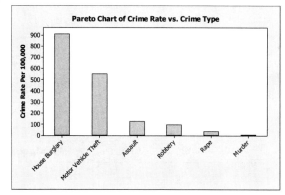

(b) Yes, but the graph would take into account only these particular crimes and would not indicate if multiple crimes occurred during the same incident.

13.

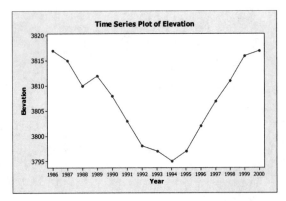

15. (a) The size of the donut hole. Make all donuts exactly the same size, with the radius of the respective holes the same as well. Data labels showing percentages for each response would also be useful.

(b) College graduates have a higher frequency of "no" responses than do those having only high school or less.

Section 2.3

1. (a) The smallest value is 47 and the largest value is 97, so we need stems 4, 5, 6, 7, 8, and 9. Use the tens digit as the stem and the ones digit as the leaf.

Longevity of Cowboys

4	7 = 47 years
4	7
5	2 7 8 8
6	1 6 6 8 8
7	0 2 2 3 3 5 6 7
8	4 4 4 5 6 6 7 9
9	0 1 1 2 3 7

(b) Yes, these cowboys certainly lived long lives, as evidenced by the high frequency of leaves for stems 7, 8, and 9 (i.e., 70-, 80-, and 90-year-olds).

3. The longest average length of stay is 11.1 days in North Dakota, and the shortest is 5.2 days in Utah. We need stems from 5 to 11. Use the digit(s) to the left of the decimal point as the stem and the digit to the right as the leaf.

Average Length of Hospital Stay

5	2 = 5.2 days
5	2 3 5 5 6 7
6	0 2 4 6 6 7 7 8 8 8 8 9 9
7	0 0 0 0 0 1 1 1 2 2 2 3 3 3 3 4 4 5 5 6 6 8
8	4 5 7
9	4 6 9
10	0 3
11	1

The distribution is skewed right.

5. **(a)** The longest time during 1961–1980 is 23 minutes (i.e., 2:23), and the shortest time is 9 minutes (2:09). We need stems 0, 1, and 2. We'll use the tens digit as the stem and the ones digit as the leaf, placing leaves 0, 1, 2, 3, and 4 on the first stem and leaves 5, 6, 7, 8, and 9 on the second stem.

Minutes Beyond 2 Hours (1961–1980)

0	9 = 9 minutes past 2 hours
0	9 9
1	0 0 2 3 3
1	5 5 6 6 7 8 8 9
2	0 2 3 3

(b) The longest time during the period 1981–2000 was 14 (2:14) and the shortest was 7 (2:07), so we'll need stems 0 and 1 only.

Minutes Beyond 2 Hours (1981–2000)

	7 = 7 minutes past 2 hours
0	7 7 7 8 8 8 8 9 9 9 9 9 9 9
1	0 0 1 1 4

(c) There were seven times under 2:15 during 1961–1980, and there were 20 times under 2:15 during 1981–2000.

7. The largest value in the data is 29.8 mg of tar per cigarette smoked, and the smallest value is 1.0. We will need stems from 1 to 29, and we will use the numbers to the right of the decimal point as the leaves.

Milligrams of Tar per Cigarette

1	0 = 1.0 mg tar
1	0
2	
3	
4	1 5
5	
6	
7	3 8
8	0 6 8
9	0
10	
11	4
12	0 4 8
13	7
14	1 5 9
15	0 1 2 8
16	0 6
17	0
.	
.	
.	
29	8

9.   The largest value in the data set is 2.03 mg of nicotine per cigarette smoked. The smallest value is 0.13. We will need stems 0, 1, and 2. We will use the number to the left of the decimal point as the stem and the first number to the right of the decimal point as the leaf. The number 2 placed to the right of the decimal point (the hundredths digit) will be truncated (not rounded).

Milligrams of Nicotine per Cigarette

0	1 = 0.1 milligram
0	1 4 4
0	5 6 6 6 7 7 7 8 8 9 9 9
1	0 0 0 0 0 0 0 1 2
1	
2	0

Chapter Review Problems

1.   **(a)**   Bar graphs, Pareto charts, pie charts
      (b)   All, but quantitative data must be categorized to use a bar graph, Pareto chart, or pie chart.

3.   Any large gaps between bars or stems might indicate potential outliers.

5.   **(a)**   Figure 2-1(a) (in the text) is essentially a bar graph with a "horizontal" axis showing years and a "vertical" axis showing miles per gallon. However, in depicting the data as a highway and showing them in perspective, the ability to correctly compare bar heights visually has been lost. For example, determining what would appear to be the bar heights by measuring from the white line on the road to the edge of the road along a line drawn from the year to its mpg value, we get the bar height for 1983 to be approximately ⅞ inch and the bar height for 1985 to be approximately 1⅜ inches (i.e., 11/8 inches). Taking the ratio of the given bar heights, we see that the bar for 1985 should be $\frac{27.5}{26} \approx 1.06$ times the length of the 1983 bar. However, the measurements show a ratio of $\frac{\frac{11}{8}}{\frac{7}{8}} = \frac{11}{7} \approx 1.60$; i.e., the 1985 bar is (visually) 1.6 times the length of the 1983 bar. Also, the years are evenly spaced numerically, but the figure shows the more recent years to be more widely spaced owing to the use of perspective.

      (b)   Figure 2-1(b) is a time-series graph showing the years on the x axis and miles per gallon on the y axis. Everything is to scale and not distorted visually by the use of perspective. It is easy to see the mpg standards for each year, and you also can see how fuel economy standards for new cars have changed over the 8 years shown (i.e., a steep increase in the early years and a leveling off in the later years).

7.   Owing to rounding, the percentages are slightly different from those in the text.

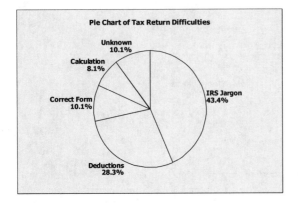

9. **(a)** The largest value is 142 mm, and the smallest value is 69. For seven classes, we need a class width of $\frac{142-69}{7} \approx 10.4$; use 11. The lower class limit of the first class is 69, and the lower class limit of the second class is $69 + 11 = 80$.

The class boundaries are the average of the upper class limit of one class and the lower class limit of the next higher class. The midpoint is the average of the class limits for that class. There are 60 data values total, so the relative frequency is the class frequency divided by 60.

Class Limits	Class Boundaries	Midpoint	Frequency	Relative Frequency	Cumulative Frequency
69–79	68.5–79.5	74	2	0.03	2
80–90	79.5–90.5	85	3	0.05	5
91–101	90.5–101.5	96	8	0.13	13
102–112	101.5–112.5	107	19	0.32	32
113–123	112.5–123.5	118	22	0.37	54
124–134	123.5–134.5	129	3	0.05	57
135–145	134.5–145.5	140	3	0.05	60

(b)

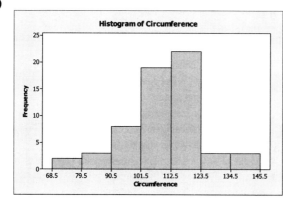

(c)

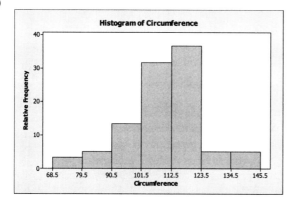

(d) This distribution is skewed left.

(e) The ogive begins on the x axis at the lower class boundary and connects dots placed at (x, y) coordinates (upper class boundary, cumulative frequency).

11. **(a)** To determine the decade that contained the most samples, count *both* rows (if shown) of leaves; recall that leaves 0–4 belong on the first line and leaves 5–9 belong on the second line when two lines per stem are used. The greatest number of leaves is found on stem 124, i.e., the 1240s (the 40s decade in the 1200s), with 40 samples.

(b) The number of samples with tree-ring dates 1200 to 1239 A.D. is $28 + 3 + 19 + 25 = 75$.

(c) The dates of the longest interval with no sample values are 1204 through 1211 A.D. This might mean that for these 8 years, the pueblo was unoccupied (thus no new or repaired structures), or that the population remained stable (no new structures needed), or that, say, weather conditions were favorable those years, so existing structures didn't need repair. If relatively few new structures were built or repaired during this period, their tree rings might have been missed during sample selection.

Chapter 3: Averages and Variation

Section 3.1

1. The middle value is the median. The most frequent value is the mode. The mean takes all values into account.

3. First add up all the data values, then divide by the number of data values.

5. The mean is $\bar{x} = \dfrac{8+2+7+2+6}{5} = 5$. The median is 6. The mode is 2.

7. The mean is $\bar{x} = \dfrac{8+2+7+2+6+5}{6} = 5$. The median is $\dfrac{5+6}{2} = 5.5$. The mode is 2.

9. (a) No, the sum of the data does not change.
   (b) No, changing the extreme data values does not affect the median.
   (c) Yes, depending on which data value occurs most frequently after the data are changed.

11. For a mound-shaped symmetrical data set, the values of the mean, median, and mode will all be equal.

13. (a) The mode is 5. The median is 4. The mean is $\bar{x} = \dfrac{2+3+4+5+5}{5} = \dfrac{19}{5} = 3.8$.
    (b) Only the mode.
    (c) All three make sense.
    (d) The mode and the median.

15. The supervisor is correct when expressing concern since at least half of the evaluations had the employee ranked as *poor* or *unacceptable*. There should also be concern because the employee evaluations are very inconsistent.

17. (a) The mode is 2. The median is 3. The mean is $\bar{x} = \dfrac{2+2+3+6+10}{5} = 4.6$.
    (b) Multiplying each value by 5 multiplies each measure of center by 5. Thus, the mode is 10, the median is 15, and the mean is 23.
    (c) In general, multiplying each value in a data set by a constant will multiply each measure of center by the same constant.
    (d) The mode is 177.8 cm. The median is 172.72 cm. The mean is 180.34 cm.

19. The mean is $\dfrac{146+152+\cdots+144}{14} \approx 167.3°\,\text{F}$. For the median, first order the data set smallest to largest. Then take the mean of the two middle values: $\dfrac{168+174}{2} = 171°\,\text{F}$. The mode is 178° F.

21. First, organize the data from smallest to largest. Then compute the mean, median, and mode.

Upper Canyon	1	1	1	2	3	3	3	3	4	6	9		

Lower Canyon	0	0	1	1	1	1	2	2	3	6	7	8	13	14

(a)   The mean is $\bar{x} = \dfrac{1+1+\cdots+9}{11} \approx 3.27$. The median is in the middle position, 3. The mode is 3.

(b)   The mean is $\bar{x} = \dfrac{0+0+\cdots+14}{14} \approx 4.21$. The median is the mean of the two middle values,

$\dfrac{2+2}{2} = 2$. The mode is 1.

(c)   The mean is lower for Upper Canyon, but the median and mode are higher.

(d)   5% of 14 is 0.7, which rounds to 1. So eliminate the smallest and largest value from the data set and compute the mean of the remaining 12 values. The trimmed mean is $\bar{x} = \dfrac{0+1+\cdots+13}{12} \approx 3.75$. This value is closer to the mean from Upper Canyon.

23. (a)   The mean is $\bar{x} = \dfrac{89+50+\cdots+130}{20} = \$136.15$. The median is $\dfrac{65+68}{2} = \$66.50$. The mode is $60.

(b)   5% of 20 data values is 1, so we remove the smallest and largest values and recalculate the mean. The trimmed mean is $\bar{x} = \dfrac{2183}{18} \approx \$121.28$. Since there are large outliers, the trimmed mean may be a more accurate measure of center than the mean.

(c)   The median is probably the best value to report, but the travel agent should also inform the clients about the high outliers as well.

25.   The weighted average is $\dfrac{\sum xw}{\sum w} = \dfrac{10(2)+20(3)+30(5)}{2+3+5} = \dfrac{230}{10} = 23$.

27.   The weighted average is $\dfrac{\sum xw}{\sum w} = \dfrac{9(2)+7(3)+6(1)+10(4)}{2+3+1+4} = 8.5$.

29.   The harmonic mean is $\dfrac{2}{\dfrac{1}{60}+\dfrac{1}{75}} \approx 66.67$ mph.

Section 3.2

1.   The mean is associated with the standard deviation.

3.   Yes. When computing the sample standard deviation, divide by $n-1$. When computing the population standard deviation, divide by n.

5. (a)   The range is $6-2=4$.

(b) $s = \sqrt{\dfrac{(2-4)^2+(3-4)^2+(4-4)^2+(5-4)^2+(6-4)^2}{5-1}} \approx 1.58$.

(c) $\sigma = \sqrt{\dfrac{(2-4)^2 + (3-4)^2 + (4-4)^2 + (5-4)^2 + (6-4)^2}{5}} \approx 1.41$.

7. For a data set in which not all the data values are equal, s will always be larger than σ. The denominator for the sample standard deviation is $n-1$, which is smaller than n, resulting in a larger sample standard deviation.

9. (a) Data set (i) has the smallest standards deviation, followed by (ii), and then (iii).

(b) The data change between data sets (i) and (ii) increased by the squared difference sum $\sum (x - \bar{x})^2$ by 10, whereas the data change between data sets (ii) and (iii) increased the squared difference sum $\sum (x - \bar{x})^2$ by only 6.

11. (a) $s = \sqrt{\dfrac{\sum (x-\bar{x})^2}{n-1}} \approx 3.61$.

(b) $s = \sqrt{\dfrac{\sum (x-\bar{x})^2}{n-1}} \approx 18.05$.

(c) Multiplying each value in a data set by a constant multiplies the standard deviation by that constant.

(d) You can simply multiply the standard deviation by 1.6. Thus, the new standard deviation is $3.1(1.6) = 4.96\ \text{km}$.

13. (a) The range is $30 - 15 = 15$.

(b) Use a calculator.

(c) $s^2 = \dfrac{\sum x^2 - \dfrac{(\sum x)^2}{n}}{n-1} = \dfrac{2568 - \dfrac{110^2}{5}}{5-1} = 37$. Then, $s = \sqrt{37} \approx 6.08$.

(d) $\bar{x} = \dfrac{110}{5} = 22$; $s^2 = \dfrac{(23-22)^2 + (17-22)^2 + (15-22)^2 + (30-22)^2 + (25-22)^2}{5-1} = \dfrac{148}{4} = 37$.

(e) To calculate the population standard deviation, divide by 5 instead of $5 - 1 = 4$. Therefore,

$\sigma^2 = \dfrac{148}{5} = 29.6$, and $\sigma = \sqrt{29.6} \approx 5.44$.

15. (a) $CV = \dfrac{\sigma}{\mu} = \dfrac{2}{20} = 10\%$.

(b) An 88.9% Chebyshev interval corresponds to ± 3 standard deviations. Therefore, the interval will be: $20 \pm 3\sigma = 20 \pm 3(2) = 20 \pm 6 = (14, 26)$.

17. (a) The range is $7.89 - 0.02 = 7.87$.

(b) Use a calculator.

(c) $\bar{x} = \dfrac{\Sigma x}{n} = \dfrac{62.11}{50} \approx 1.24$, $s^2 = \dfrac{\Sigma x^2 - \frac{(\Sigma x)^2}{n}}{n-1} = \dfrac{164.23 - \frac{(62.11)^2}{50}}{50 - 1} \approx 1.78$, $s \approx \sqrt{1.78} \approx 1.33$.

(d) $CV = \dfrac{1.33}{1.24} \approx 107.3\%$. The standard deviation of the time to failure is just slightly larger than the average time to failure.

19. (a)   Students verify results with a calculator.

x	x^2	y	y^2
13.20	174.24	11.85	140.42
5.60	31.36	15.25	232.56
19.80	392.04	21.30	453.69
15.05	226.50	17.30	299.29
21.40	457.96	27.50	756.25
17.25	297.56	10.35	107.12
27.45	753.50	14.90	222.01
16.95	287.30	48.70	2371.69
23.90	571.21	25.40	645.16
32.40	1049.76	25.95	673.40
40.75	1660.56	57.60	3317.76
5.10	26.01	34.35	1179.92
17.75	315.06	38.80	1505.44
28.35	803.72	41.00	1681.00
		31.25	976.56

(b) $\bar{x} = \dfrac{\Sigma x}{n} = \dfrac{245}{5} = 49$, $s = \sqrt{\dfrac{\Sigma x^2 - \frac{(\Sigma x)^2}{n}}{n-1}} = \sqrt{\dfrac{14,755 - \frac{(245)^2}{5}}{5-1}} \approx 26.22$, $s^2 = 26.22^2 \approx 687.49$

(c) $\bar{y} = \dfrac{\Sigma y}{n} = \dfrac{224}{5} = 44.8$, $s = \sqrt{\dfrac{\Sigma y^2 - \frac{(\Sigma y)^2}{n}}{n-1}} = \sqrt{\dfrac{12,070 - \frac{(224)^2}{5}}{5-1}} \approx 22.55$, $s^2 = 22.55^2 \approx 508.50$

(d) Mallard nest: $CV = \dfrac{s}{x} \cdot 100 = \dfrac{26.22}{49} \cdot 100 \approx 53.5\%$

Canada Goose nest: $CV = \dfrac{s}{y} \cdot 100 = \dfrac{22.55}{44.8} \cdot 100 \approx 50.3\%$

The CV gives the ratio of the standard deviation to the mean. With respect to their means, the variation for the mallards is slightly higher than the variation for the Canada geese.

21. $CV = \dfrac{s}{\bar{x}} \cdot 100$, $\quad s = \dfrac{\bar{x} \cdot CV}{100} = \dfrac{2.2(1.5)}{100} = 0.033$.

23.

Class	f	x	xf	$x - \bar{x}$	$(x - \bar{x})^2$	$(x - \bar{x})^2 f$
21–30	260	25.5	6630	−10.3	106.09	27,583.4
31–40	348	35.5	12,354	−0.3	0.09	31.3
41 and over	287	45.5	13,058.5	9.7	94.09	27,003.8
	$n = \sum f = 895$		$\sum xf = 32{,}042.5$			$\sum (x - \bar{x})^2 f = 54{,}619$

$\bar{x} = \dfrac{\sum xf}{n} = \dfrac{32{,}042.5}{895} \approx 35.80$

$s^2 = \dfrac{\sum (x - \bar{x})^2 \cdot f}{n-1} = \dfrac{54{,}619}{894} \approx 61.1$

$s = \sqrt{61.1} \approx 7.82$

25.

Class	f	x	xf	$x - \bar{x}$	$(x - \bar{x})^2$	$(x - \bar{x})^2 f$
8.6–12.5	15	10.55	158.25	−5.05	25.502	382.537
12.6–16.5	20	14.55	291.00	−1.05	1.102	22.050
16.6–20.5	5	18.55	92.75	2.95	8.703	43.513
20.6–24.5	7	22.55	157.85	6.95	48.303	338.118
24.6–28.5	3	26.55	79.65	10.95	19.903	359.708
	$n = \sum f = 50$		$\sum xf = 779.5$			$\sum (x - \bar{x})^2 f = 1{,}145.9$

$\bar{x} = \dfrac{\sum xf}{n} = \dfrac{779.5}{50} \approx 15.6$

$s^2 = \dfrac{\sum (x - \bar{x})^2 f}{n-1} = \dfrac{1{,}145.9}{49} \approx 23.4$

$s = \sqrt{23.4} \approx 4.8$

27.

$$\sum(x-\bar{x})^2 = \sum\left(x^2 - 2x\bar{x} + \bar{x}^2\right) = \sum x^2 - \sum 2x\bar{x} + \sum \bar{x}^2 =$$

$$\sum x^2 - 2\bar{x}\sum x + n\bar{x}^2 = \sum x^2 - 2\bar{x}n\bar{x} + n\bar{x}^2 =$$

$$\sum x^2 - 2n\bar{x}^2 + n\bar{x}^2 = \sum x^2 - n\bar{x}^2 = \sum x^2 - n\left(\frac{\sum x}{n}\right)^2 =$$

$$\sum x^2 - \frac{\left(\sum x\right)^2}{n}$$

29. (a)

$$n_1 = \left[\frac{1525(2.2)}{1525(2.2)+917(1.4)+2890(3.3)}\right]250 \approx 59.168 \approx 59$$

$$n_2 = \left[\frac{917(1.4)}{1525(2.2)+917(1.4)+2890(3.3)}\right]250 \approx 22.641 \approx 23$$

$$n_3 = \left[\frac{2890(3.3)}{1525(2.2)+917(1.4)+2890(3.3)}\right]250 \approx 168.192 \approx 168$$

(b) $\mu \approx \dfrac{59}{250}6.2 + \dfrac{23}{250}3.1 + \dfrac{168}{250}8.5 \approx 7.46$

Section 3.3

1. 82% of the scores were at or below her score. $100\% - 82\% = 18\%$ of the scores were above her score.

3. No, the score 82 might have a percentile rank less than 70. Raw scores are not necessarily equal to percentile scores.

5. (a) The low is 2, $Q_1 = 5$, the median is 7, $Q_3 = 8.5$, and the high is 10.

(b) $IQR = 8.5 - 5 = 3.5$.

(c)

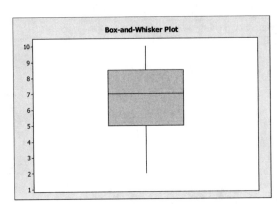

7. Order the data from smallest to largest.

Lowest value = 2
Highest value = 42

There are 20 data values.

$$\text{Median} = \frac{23+23}{2} = 23$$

There are 10 values less than the Q_2 position and 10 values greater than the Q_2 position.

$$Q_1 = \frac{8+11}{2} = 9.5$$

$$Q_3 = \frac{28+29}{2} = 28.5$$

$$IQR = Q_3 - Q_1 = 28.5 - 9.5 = 19$$

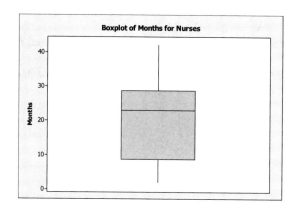

9. (a) Lowest value = 17
Highest value = 38

There are 50 data values.

$$\text{Median} = \frac{24+24}{2} = 24$$

There are 25 values above and 25 values below the Q_2 position.

$$Q_1 = 22$$
$$Q_3 = 27$$
$$IQR = 27 - 22 = 5$$

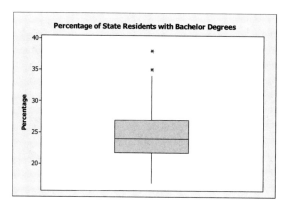

(b) 26% is in the third quartile because it is between the median and Q_3.

11. (a) California has the lowest premium, and Pennsylvania has the highest.
  (b) Pennsylvania has the highest median premium.
  (c) California has the smallest range, and Texas has the smallest *IQR*.
  (d) The smallest *IQR* will be Texas. The largest *IQR* will be Pennsylvania.
  For figure (a), $IQR = 3{,}652 - 2{,}758 = 894$
  For figure (b), $IQR = 5{,}801 - 4{,}326 = 1{,}475$
  For figure (c), $IQR = 3{,}966 - 2{,}801 = 1{,}165$
  Therefore, figure (a) is Texas and figure (b) is Pennsylvania. By elimination, figure (c) is California.

Chapter Review Problems

1. (a) The variance and the standard deviation
  (b) Box-and-whisker plot

3. (a) For both data sets, the mean is 20.
  Also, for both data sets, the range = maximum − minimum = 31 − 7 = 24.
  (b) Data set C1 seems more symmetric because the mean equals the median and the median is centered in the interquartile range.
  (c) For C1, $IQR = 25 - 15 = 10$. For C2, $IQR = 22 - 20 = 2$. Thus, for C1, the middle 50% of the data have a range of 10, whereas for C2, the middle 50% of the data have a smaller range of 2.

5. (a)   Lowest value = 31
  Highest value = 68

  There are 60 data values.

  $$\text{Median} = \frac{45 + 45}{2} = 45$$

  There are 30 values above and 30 values below the Q_2 position.

  $$Q_1 = \frac{40 + 40}{2} = 40$$

  $$Q_3 = \frac{52 + 53}{2} = 52.5$$

  $$IQR = 52.5 - 40 = 12.5$$

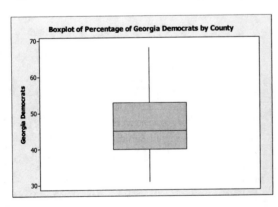

(b) Class width = 8

Class	x Midpoint	f	xf	$x^2 f$
31–38	34.5	11	379.5	13,092.8
39–46	42.5	24	1020	43,350.0
47–54	50.5	15	757.5	38,253.8
55–62	58.5	7	409.5	23,955.8
63–70	66.5	3	199.5	13,266.8
		$n = \sum f = 60$	$\sum xf = 2,766$	$\sum x^2 f = 131,919$

$$\bar{x} = \frac{\sum xf}{n} = \frac{2,766}{60} = 46.1$$

$$s = \sqrt{\frac{\sum x^2 f - \frac{(\sum xf)^2}{n}}{n-1}} = \sqrt{\frac{131,919 - \frac{(2,766)^2}{60}}{59}} = \sqrt{\frac{4,406.4}{59}} \approx 8.64$$

$$\bar{x} - 2s = 46.1 - 2(8.64) = 28.82$$
$$\bar{x} + 2s = 46.1 + 2(8.64) = 63.38$$

We expect at least 75% of the counties in Georgia to have between 28.82% and 63.38% Democrats.

(c) $\bar{x} = 46.15$, $s \approx 8.63$

7. Mean weight $= \dfrac{2,500}{16} = 156.25$

The mean weight is 156.25 lb.

9. **(a)** A college degree does not guarantee an increase of 83.4% in earnings compared with a high-school diploma. This statement is based on averages.
   (b) We compute as follows:

   $$\bar{x} - 2s = \$51,206 - 2(\$8,500) = \$34,206$$
   $$\bar{x} + 2s = \$51,206 + 2(\$8,500) = \$68,206$$

   (c)

   $$\bar{x} = \frac{(0.46)(4,500) + (0.21)(7,500) + (0.07)(12,000) + (0.08)(18,000) + (0.09)(24,000) + (0.09)(31,000)}{0.46 + 0.21 + 0.07 + 0.08 + 0.09 + 0.09} =$$

   $$\bar{x} = \$10,875$$

11. Weighted average $= \dfrac{\sum xw}{\sum w} = \dfrac{5(2) + 8(3) + 7(3) + 9(5) + 7(3)}{2 + 3 + 3 + 5 + 3} = \dfrac{121}{16} \approx 7.56$

Cumulative Review Problems Chapters 1, 2, 3

1. (a) Median, percentile
   (b) Mean, variance, standard deviation

3. (a) Same
   (b) Set B has higher mean.
   (c) Set B has higher standard deviation.
   (d) Set B has much longer whisker beyond Q_3.

5. One could assign a consecutive number to each well in West Texas and then use a random-number table or a computer package to draw the simple random sample.

7. Use the ones digit for the stem and the tenths decimal for the leaves. Split each stem into five rows. Here, 7 0 = 7.0.

   7 000000001111111111
   7 222222222233333333333
   7 44444444455555555
   7 666666666777777
   7 8888899999
   8 01111111
   8 2222222
   8 45
   8 67
   8 88

9. To draw the ogive, the vertical axis is labeled with relative frequency, and the horizontal axis is labeled with the upper class boundaries. Draw a dot at the minimum class boundary and zero, and then draw a dot at each upper class boundary and the corresponding cumulative frequency. Connect the dots.

11. (a) The students can verify the figures using a calculator or a statistics package.
    (b)

$$s^2 = \frac{\sum(x-\bar{x})^2}{n-1} = 0.1984$$

$$s = \sqrt{s^2} = \sqrt{0.1984} = 0.4454$$

$$CV = \frac{s}{\bar{x}} = \frac{0.4454}{7.58} = 0.59 = 5.9\%$$

The sample variance is only 5.9% of the mean. This appears to be small.

13. We know the minimum value is 7.0, the maximum value is 8.8, and the median is 7.5. Using Minitab, we find that $Q_1 = 7.2$ and $Q_3 = 7.9$. Thus $IQR = 7.9 - 7.2 = 0.7$.

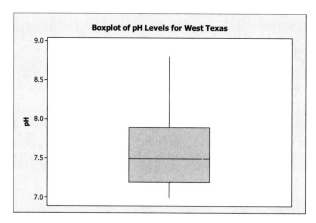

15. 87.2% of the wells have a pH of less than 8.15. 57.8% of the wells could be used for the irrigation. Here, 57.8% = 31.4% + 17.6% + 8.8%.

17. Half the wells are found to have a pH between 7.2 and 7.9. There is skewness toward the high values, with half the wells having a pH between 7.5 and 8.8. The boxplot and the histogram are consistent because both show the distribution to be right skewed.

Chapter 4: Elementary Probability Theory

Section 4.1

1. Equally likely outcomes, relative frequency, intuition.

3. **(a)** The probability of a certain event is 1.
   (b) The probability of an impossible event is 0.

5. No, the probability was stated for drivers in the age range from 18 to 24. We have no information for other age groups. Other age groups may not behave the same way as the 18 to 24 year olds.

7. $P(\text{Very Great Prestige}) = \dfrac{627}{1010} \approx 0.62$.

9. You are not certain to make money on this investment as there is a 3% chance you could lose your entire investment. It is likely, based on the given information, to double your money, but there are risks involved.

11. **(a)** MMM, MMF, MFM, MFF, FMM, FMF, FFM, FFF.
    (b) $P(\text{MMM}) = \dfrac{1}{8}$ and $P(\text{At least one child is female}) = 1 - P(\text{MMM}) = 1 - \dfrac{1}{8} = \dfrac{7}{8}$.

13. No. The probability of throwing tails on the second toss is 0.5 regardless of the outcome on the first toss.

15. The resulting relative frequency can be used as an estimate of the true probability of all Americans who can wiggle their ears.

17. **(a)** $P(\text{no similar preferences}) = P(0) = \dfrac{15}{375}$, $P(1) = \dfrac{71}{375}$, $P(2) = \dfrac{124}{375}$, $P(3) = \dfrac{131}{375}$, $P(4) = \dfrac{34}{375}$
    (b) $\dfrac{15 + 71 + 124 + 131 + 34}{375} = \dfrac{375}{375} = 1$, yes
    Personality types were classified into four main preferences; all possible numbers of shared preferences were considered. The sample space is 0, 1, 2, 3, and 4 shared preferences.

19. **(a)** $P(\text{best idea 6 A.M.–12 noon}) = \dfrac{290}{966} \approx 0.30$

    $P(\text{best idea 12 noon–6 P.M.}) = \dfrac{135}{966} \approx 0.14$

    $P(\text{best idea 6 P.M.–12 midnight}) \dfrac{319}{966} \approx 0.33$

    $P(\text{best idea from 12 midnight to 6 A.M.}) = \dfrac{222}{966} \approx 0.23$

    (b) The probabilities add up to 1. They should add up to 1 provided that the intervals do not overlap and each inventor chose only one interval. The sample space is the set of four time intervals.

21. **(a)** *Given: Odds in favor of A are n:m* $\left(\text{i.e., } \dfrac{n}{m} \right)$.

Show: $P(A) = \dfrac{n}{m+n}$

Proof: Odds in favor of A are $\dfrac{P(A)}{P(\text{not } A)}$ by definition

$$P(\text{not } A) = 1 - P(A) \qquad \text{complementary events}$$

$$\frac{n}{m} = \frac{P(A)}{P(\text{not } A)} = \frac{P(A)}{1 - P(A)} \qquad \text{substitution}$$

$$n[1 - P(A)] = m[P(A)] \qquad \text{cross multiply}$$

$$n - n[P(A)] = m[P(A)]$$

$$n = n[P(A)] + m[P(A)]$$

$$n = (n + m)[P(A)]$$

So $\dfrac{n}{n+m} = P(A)$, as was to be shown.

(b) Odds of a successful call are 2 to 15. Now 2 to 15 can be written as 2:15 or $\dfrac{2}{15}$.

From part **(a):** if the odds are 2:15 (let $n = 2$, $m = 15$), then $P(\text{sale}) = \dfrac{n}{n+m} = \dfrac{2}{2+15} = \dfrac{2}{17} \approx 0.118$.

(c) Odds of free throw are 3 to 5, i.e., 3:5.

Let $n = 3$ and $m = 5$ here; then, from part (a):

$$P(\text{free throw}) = \frac{n}{n+m} = \frac{3}{3+5} = \frac{3}{8} = 0.375$$

23. One approach is to make a table showing the information about the 127 people who walked by the store.

	Buy	Did Not Buy	Row Total
Came into the store	25	$58 - 25 = 33$	58
Did not come in	0	69	$127 - 58 = 69$
Column total	25	102	127

If 58 came in, 69 didn't; 25 of the 58 bought something, so 33 came in but didn't buy anything. Those who did not come in couldn't buy anything. The row entries must sum to the row totals, the column entries must sum to the column totals, and the row totals, as well as the column totals, must sum to the overall total, i.e., the 127 people who walked by the store. Also, the four inner cells must sum to the overall total:

$25 + 33 + 0 + 69 = 127$.

This kind of problem relies on formula (2), $P(\text{event } A) = \dfrac{\text{number outcomes favorable to } A}{\text{total number of outcomes}}$.

(a) $P(\text{Enter the store}) = \dfrac{58}{127} \approx 0.46$. Here, we divide by 127 people.

(b) $P(\text{Buy something after entering the store}) = \dfrac{25}{58} \approx 0.43$. Here, we divide by 58 people.

(c) $P(\text{Enter and buy}) = \dfrac{58}{127} \times \dfrac{25}{58} = \dfrac{25}{127} \approx 0.20$

Or similarly, read from the table that 25 people both entered and bought something. Divide this by the total number of people, namely, 127.

(d) $P(\text{Buy nothing after entering the store}) = \dfrac{33}{58} \approx 0.57$. Here, we divide by 58 people.

Section 4.2

1. No. Mutually exclusive events cannot occur at the same time.

3. **(a)** $P(A \text{ or } B) = P(A) + P(B) = 0.3 + 0.4 = 0.7$.

   (b) $P(A \text{ or } B) = P(A) + P(B) - P(A \text{ and } B) = 0.3 + 0.4 - 0.1 = 0.6$.

5. **(a)** $P(A \text{ and } B) = P(A)P(B) = 0.2(0.4) = 0.08$.

   (b) $P(A \text{ and } B) = P(B)P(A \mid B) = 0.4(0.1) = 0.04$.

7. **(a)** $P(A \text{ and } B) = P(B)P(A \mid B) = 0.5(0.3) = 0.15$.

   (b) $P(A \text{ or } B) = P(A) + P(B) - P(A \text{ and } B) = 0.2 + 0.5 - 0.15 = 0.55$.

9. *P(A and B)* is the probability that both events *A* and *B* occur. It cannot exceed the probability that either event occurs. When the assigned probabilities are used to get P(*A* | *B*), the result exceeds 1.

11. **(a)** Event *A* cannot occur if event *B* has occurred. Therefore, $P(A \mid B) = 0$.

    (b) Since we are told that $P(A) \neq 0$, and we have determined that $P(A \mid B) = 0$, we can deduce that $P(A) \neq P(A \mid B)$. Therefore, events *A* and *B* are not independent.

13. **(a)** $P(A \text{ and } B)$ **(b)** $P(B \mid A)$ **(c)** $P(A^c \mid B)$ **(d)** $P(A \text{ or } B)$ **(e)** $P(A \text{ or } B^c)$

15. **(a)** Green and blue are mutually exclusive because each M&M candy is only one color.
    $P(\text{green or blue}) = P(\text{green}) + P(\text{blue}) = 10\% + 10\% = 20\% = 0.20$.

    (b) Yellow and red are mutually exclusive once again because each candy is only one color.
    $P(\text{yellow or red}) = P(\text{yellow}) + P(\text{red}) = 20\% + 20\% = 40\% = 0.40$.

    (c) Use the complementary event.
    $P(\text{not purple}) = 1 - P(\text{purple}) = 1 - 0.20 = 0.80 = 80\%$

17. **(a)** Yes, the outcome of the red die does not influence the outcome of the green die.

    (b) $P(5 \text{ on green and } 3 \text{ on red}) = P(5 \text{ on green}) \cdot P(3 \text{ on red}) = \left(\dfrac{1}{6}\right)\left(\dfrac{1}{6}\right) = \dfrac{1}{36} \approx 0.028$.

    (c) $P(3 \text{ on green and } 5 \text{ on red}) = P(3 \text{ on green}) \cdot P(5 \text{ on red}) = \left(\dfrac{1}{6}\right)\left(\dfrac{1}{6}\right) = \dfrac{1}{36} \approx 0.028$

    (d) $P[(5 \text{ on green and } 3 \text{ on red}) \text{ or } (3 \text{ on green and } 5 \text{ on red})]$

    $= P(5 \text{ on green and } 3 \text{ on red}) + P(3 \text{ on green and } 5 \text{ on red})$

    $= \dfrac{1}{36} + \dfrac{1}{36} = \dfrac{2}{36} = \dfrac{1}{18} \approx 0.056$ (because they are mutually exclusive outcomes).

19. (a) We can obtain a sum of 6 as follows:

$$1 + 5 = 6$$
$$2 + 4 = 6$$
$$3 + 3 = 6$$
$$4 + 2 = 6$$
$$5 + 1 = 6$$

$P(\text{sum} = 6) = P[(1, 5) \text{ or } (2, 4) \text{ or } (3 \text{ on red}, 3 \text{ on green}) \text{ or } (4, 2) \text{ or } (5, 1)]$

$\qquad = P(1, 5) + P(2, 4) + P(3, 3) + P(4, 2) + P(5, 1)$

because the (red, green) outcomes are mutually exclusive

$\qquad = \left(\dfrac{1}{6}\right)\left(\dfrac{1}{6}\right) + \left(\dfrac{1}{6}\right)\left(\dfrac{1}{6}\right) + \left(\dfrac{1}{6}\right)\left(\dfrac{1}{6}\right) + \left(\dfrac{1}{6}\right)\left(\dfrac{1}{6}\right) + \left(\dfrac{1}{6}\right)\left(\dfrac{1}{6}\right)$

because the red die outcome is independent of the green die outcome

$\qquad = \dfrac{1}{36} + \dfrac{1}{36} + \dfrac{1}{36} + \dfrac{1}{36} + \dfrac{1}{36} = \dfrac{5}{36}$

(b) We can obtain a sum of 4 as follows:

$$1 + 3 = 4$$
$$2 + 2 = 4$$
$$3 + 1 = 4$$

$P(\text{sum is } 4) = P[(1, 3) \text{ or } (2, 2) \text{ or } (3, 1)]$

$\qquad = P(1, 3) + P(2, 2) + P(3, 1)$

because the (red, green) outcomes are mutually exclusive

$\qquad = \left(\dfrac{1}{6}\right)\left(\dfrac{1}{6}\right) + \left(\dfrac{1}{6}\right)\left(\dfrac{1}{6}\right) + \left(\dfrac{1}{6}\right)\left(\dfrac{1}{6}\right)$

because the red die outcome is independent of the green die outcome

$\qquad = \dfrac{1}{36} + \dfrac{1}{36} + \dfrac{1}{36} = \dfrac{3}{36} = \dfrac{1}{12}$

(c) You cannot roll a sum of 6 and a sum of 4 at the same time. These are mutually exclusive events.

$P(\text{sum of 6 or 4}) = P(\text{sum of 6}) + P(\text{sum of 4}) = \dfrac{5}{36} + \dfrac{3}{36} = \dfrac{8}{36} = \dfrac{2}{9}$

21. (a) No, the draws are not independent. The key idea is "without replacement" because the probability of the second card drawn depends on the first card drawn. Let the card draws be represented by an (x, y) ordered pair. For example, $(K, 6)$ means the first card drawn was a king and the second card drawn was a 6. Here the order of the cards is important.

(b) $P(\text{ace on first draw and king on second draw}) = P(\text{ace, king}) = \left(\dfrac{4}{52}\right)\left(\dfrac{4}{51}\right) = \dfrac{16}{2,652} = \dfrac{4}{663}$

There are four aces and four kings in the deck. Once the first card is drawn and not replaced, there are only 51 cards left to draw from, but all the kings are available.

(c) $P(\text{king, ace}) = \left(\dfrac{4}{52}\right)\left(\dfrac{4}{51}\right) = \dfrac{16}{2652} = \dfrac{4}{663}$

(d) $P(\text{ace and king in either order}) = P[(\text{ace, king}) \text{ or } (\text{king, ace})] = P(\text{ace, king}) + P(\text{king, ace})$

$\qquad = \dfrac{16}{2,652} + \dfrac{16}{2,652} = \dfrac{32}{2,652} = \dfrac{8}{663}$

23. **(a)** Yes, the draws are independent. The key idea is "with replacement." When the first card drawn is replaced, the sample space is the same for the second card as it was for the first card. In fact, it is possible to draw the same card twice. Let the card draws be represented by an (x, y) ordered pair; for example, (K, 6) means a king was drawn, replaced, and then the second card, a 6, was drawn.

(b) $P(A, K) = P(A) \cdot P(K)$ because they are independent.

$$= \left(\frac{4}{52}\right)\left(\frac{4}{52}\right) = \frac{16}{2,704} = \frac{1}{169}$$

(c) $P(K, A) = P(K) \cdot P(A)$ because they are independent.

$$= \left(\frac{4}{52}\right)\left(\frac{4}{52}\right) = \frac{16}{2,704} = \frac{1}{169}$$

(d) $P[(A, K) \text{ or } (K, A)] = P(A, K) + P(K, A)$ because the two outcomes are mutually exclusive.

$$= \frac{1}{169} + \frac{1}{169} = \frac{2}{169}$$

25. **(a)** $P(6 \text{ years old or older}) = 27\% + 14\% + 22\% = 63\%$
    (b) $P(12 \text{ years old or younger}) = 1 - P(13 \text{ years old or older}) = 100\% - 22\% = 78\%$
    (c) $P(\text{Between 6 and 12 years old}) = 27\% + 14\% = 41\%$
    (d) $P(\text{Between 2 and 9 years old}) = 22\% + 27\% = 49\%$

The 13-and-older category may include children up to 17 or 18 years old. This is a larger category.

27. Let T denote "telling the truth." Let L denote "machine catches a person lying." We are given the following probabilities:

$P(L \mid T^c) = 0.72 \qquad P(L \mid T) = 0.07$

(a) Given $P(T) = 0.90$. Then

$P(T \text{ and } L) = P(T) \times P(L \mid T) = (0.90) \times (0.07) = 0.063$

(b) Given $P(T^c) = 0.10$. Then

$P(T^c \text{ and } L) = P(T^c) \times P(L \mid T^c) = (0.10) \times (0.72) = 0.072$

(c) Given $P(T) = P(T^c) = 0.50$. Then

$P(T \text{ and } L) = (0.50) \times (0.07) = 0.035$
$P(T^c \text{ and } L) = (0.50) \times (0.72) = 0.36$

(d) Given $P(T) = 0.15$ and $P(T^c) = 0.85$. Then

$P(T \text{ and } L) = (0.15) \times (0.07) = 0.0105$
$P(T^c \text{ and } L) = (0.85) \times (0.72) = 0.612$

29. **(a)** $P(S) = \dfrac{686}{1,160}$

$P(S \mid A) = \dfrac{270}{580}$

$$P(S \mid Pa) = \frac{416}{580}$$

(b) No, they are not independent. $P(S \mid Pa) \neq P(S)$ based on the previous part.
(c) $P(A \text{ and } S) = 270/1{,}160$ using the table.

$P(Pa \text{ and } S) = 416/1{,}160$ using the table.

(d) $P(N) = \dfrac{474}{1{,}160}$

$P(N \mid A) = \dfrac{310}{580}$

(e) No, they are not independent. $P(N \mid A) \neq P(N)$ based on the preceding part.

(f) $P(A \text{ or } S) = P(A) + P(S) - P(A \text{ and } S)$
$$= \frac{580}{1{,}160} + \frac{686}{1{,}160} - \frac{270}{1{,}160} = \frac{996}{1{,}160}$$

31. Let C denote the presence of the condition and *not C* denote absence of the condition.

(a) $P(+ \mid C) = \dfrac{72}{154}$

(b) $P(- \mid C) = \dfrac{82}{154}$

(c) $P(- \mid \text{not } C) = \dfrac{79}{116}$

(d) $P(+ \mid \text{not } C) = \dfrac{37}{116}$

(e) $P(C \text{ and } +) = P(C) \times P(+ \mid C) = \left(\dfrac{154}{270}\right)\left(\dfrac{72}{154}\right) = \dfrac{72}{270}$

(f) $P(C \text{ and } -) = P(C) \times P(- \mid C) = \left(\dfrac{154}{270}\right)\left(\dfrac{82}{154}\right) = \dfrac{82}{270}$

33. *Given:* Let A be the event that a new store grosses > \$940,000 in year 1; then A^c is the event the new store grosses $\leq$ \$940,000 the first year.

Let B be the event that the store grosses > \$940,000 in the second year; then B^c is the event the store grosses $\leq$ \$940,000 in the second year of operation.

2-Year Results	Translations
A and B	Profitable both years
A and B^c	Profitable first but not second year
A^c and B	Profitable second but not first year
A^c and B^c	Not profitable either year

$P(A) = 65\%$ (show profit in first year)
$P(A^c) = 35\%$
$P(B) = 71\%$ (show profit in second year)
$P(B^c) = 29\%$
$P(\text{close}) = P(A^c \text{ and } B^c)$
$P(B \mid A) = 87\%$

    (a) $P(A) = 65\% = 0.65$
    (b) $P(B) = 71\% = 0.71$
    (c) $P(B \mid A) = 87\% = 0.87$
    (d) $P(A \text{ and } B) = P(A) \times P(B \mid A) = (0.65)(0.87) = 0.5655 \approx 0.57$
    (e) $P(A \text{ or } B) = P(A) + P(B) - P(A \text{ and } B) = 0.65 + 0.71 - 0.57 = 0.79$
    (f) $P(\text{not closed}) = P(\text{show a profit in year 1 or year 2 or both}) = 0.79$
        $P(\text{closed}) = 1 - P(\text{not closed}) = 1 - 0.79 = 0.21$

35. Let *TB* denote that the person has tuberculosis.
    Let + denote the test for tuberculosis indicates the presence of the disease.
    Let – denote the test for tuberculosis indicates the absence of the disease.

    We are given the following probabilities:
    $P(+ \mid TB) = 0.82$ (sensitivity of the test)
    $P(+ \mid TB^c) = 0.09$ (false-positive rate)
    $P(TB) = 0.04$

    (a) $P(TB \text{ and } +) = P(+ \mid TB) \times P(TB) = (0.82) \times (0.04) = 0.0328$

    (b) $P(TB^c) = 1 - P(TB) = 1 - 0.04 = 0.96$

    (c) $P(TB^c \text{ and } +) = P(+ \mid TB^c) \times P(TB^c) = (0.09) \times (0.96) = 0.0864$

37. True. These are complementary events and share no outcomes in common.

39. False. If event A^c has occurred, then event A cannot occur.

41. True. $P(A \mid B) = \dfrac{P(A \text{ and } B)}{P(B)} \geq P(A \text{ and } B)$ since $0 < P(B) \leq 1$.

43. True. All the outcomes in event A and B are also in event A.

45. True. All the outcomes in event A^c *and* B^c are also in event A^c.

47. False. See problem 11.

49. True. Since $P(A \text{ and } B) = P(A)P(B) = 0$, at least one of $P(A)$ or $P(B)$ must be zero.

51. True. All events under the condition "given B" are included in A or A^c.

Section 4.3

1. The permutations rule counts the number of different arrangements, or r items out of n distinct items. Here, the ordering matters. The combinations rule counts the number of groups of r items out of n distinct items. Here, the ordering does not matter. For a permutation, ABC is different from ACB. For a combination, ABC and ACB are the same item. The number of permutations is larger than the number of combinations.

3. **(a)** Use the combinations rule because we are concerned only with the groups of size five.
    (b) Use the permutations rules because we are concerned with the number of different arrangements of size five.

5.   **(a)**

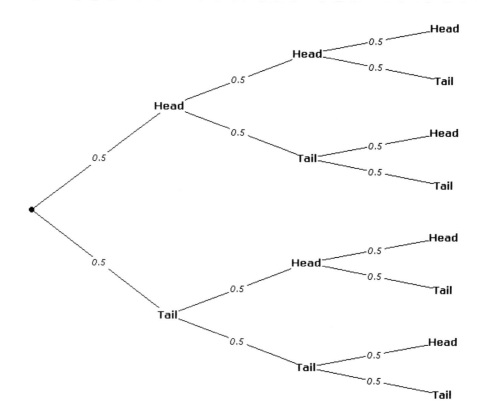

(b)   HHT, HTH, THH. There are three outcomes.

(c)   There are eight possible outcomes, and three outcomes have exactly two heads. $\dfrac{3}{8}$.

7.   (a)

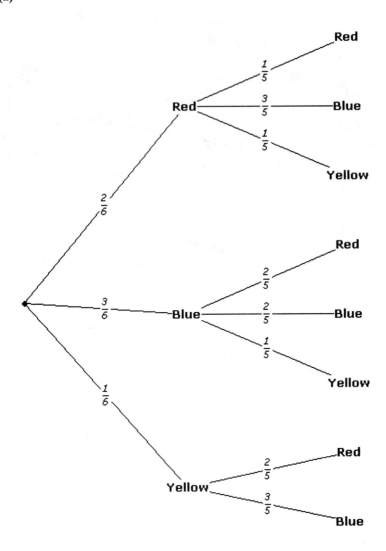

(b) Let $P(x, y)$ be the probability of choosing an x-colored ball on the first draw and a y-colored ball on the second draw. Notice that the probabilities add to 1.

$$P(B, R) = \left(\frac{3}{6}\right)\left(\frac{2}{5}\right) = \frac{6}{30} = \frac{1}{5}$$

$$P(R, R) = \left(\frac{2}{6}\right)\left(\frac{1}{5}\right) = \frac{2}{30} = \frac{1}{15} \qquad P(B, B) = \left(\frac{3}{6}\right)\left(\frac{2}{5}\right) = \frac{6}{30} = \frac{1}{5}$$

$$P(R, B) = \left(\frac{2}{6}\right)\left(\frac{3}{5}\right) = \frac{6}{30} = \frac{1}{5} \qquad P(B, Y) = \left(\frac{3}{6}\right)\left(\frac{1}{5}\right) = \frac{3}{30} = \frac{1}{10}$$

$$P(R, Y) = \left(\frac{2}{6}\right)\left(\frac{1}{5}\right) = \frac{2}{30} = \frac{1}{15} \qquad P(Y, R) = \left(\frac{1}{6}\right)\left(\frac{2}{5}\right) = \frac{2}{30} = \frac{1}{15}$$

$$P(Y, B) = \left(\frac{1}{6}\right)\left(\frac{3}{5}\right) = \frac{3}{30} = \frac{1}{10}$$

9. Using the provided hint, we multiply. There are $4 \times 3 \times 2 \times 1 = 4! = 24$ possible wiring configurations.

11. There are four fertilizers, three temperature zones for each fertilizer, and three water treatments for every fertilizer–temperature zone combination. She needs to test $4 \times 3 \times 3 = 36$ plots.

Problems 13 and 15 deal with permutations.
Use $P_{n,r} = \dfrac{n!}{(n-r)!}$ to count the number of ways r objects can be selected from n objects when ordering matters.

13. $P_{5,2} : n = 5, r = 2$ $\qquad P_{5,2} = \dfrac{5!}{(5-2)!} = \dfrac{5 \cdot 4 \cdot 3 \cdot 2 \cdot 1}{3!} = 20$

15. $P_{7,7} : n = r = 7$ $\qquad P_{7,7} = \dfrac{7!}{(7-7)!} = \dfrac{7!}{0!} = 7! = 5,040$ (recall $0! = 1$)

In general, $P_{n,n} = \dfrac{n!}{(n-n)!} = \dfrac{n!}{0!} = \dfrac{n!}{1} = n!$.

Problems 17 and 19 deal with combinations.
Use $C_{n,r} = \dfrac{n!}{r!(n-r)!}$ to count the number of ways r objects can be selected from n objects when ordering is irrelevant.

17. $C_{5,2} : n = 5, r = 2$ $\qquad C_{5,2} = \dfrac{5!}{2!(5-2)!} = \dfrac{5!}{2!3!} = \dfrac{5 \cdot 4 \cdot 3 \cdot 2 \cdot 1}{2 \cdot 1 \cdot 3 \cdot 2 \cdot 1} = \dfrac{20}{2} = 10$

19. $C_{7,7} : n = r = 7$ $\qquad C_{7,7} = \dfrac{7!}{7!(7-7)!} = \dfrac{7!}{7!0!} = \dfrac{7!}{7!(1)} = 1$ (recall $0! = 1$)

In general, $C_{n,n} = \dfrac{n!}{n!(n-n)!} = \dfrac{n!}{n!0!} = \dfrac{n!}{n!(1)} = 1$. There is only one way to choose n objects without regard to order.

21. Since the order matters (first is day supervisor, second is night supervisor, and third is coordinator), this is a permutation of 15 nurse candidates to fill three positions.

$$P_{15,3} = \dfrac{15!}{(15-3)!} = \dfrac{15!}{12!} = \dfrac{15 \cdot 14 \cdot 13 \cdot 12!}{12!} = 2,730$$

23. Order matters because the resulting sequence determines who wins first, second, and third place.

$$P_{5,3} = \dfrac{5!}{(5-3)!} = \dfrac{5!}{2!} = \dfrac{120}{2} = 60$$

25. The order of trainee selection is irrelevant, so use the combinations method.

$$C_{15,5} = \dfrac{15!}{5!(15-5)!} = \dfrac{15!}{5!10!} = \dfrac{15 \cdot 14 \cdot 13 \cdot 12 \cdot 11 \cdot 10!}{5!10!} = \dfrac{15 \cdot 14 \cdot 13 \cdot 12 \cdot 11}{5 \cdot 4 \cdot 3 \cdot 2 \cdot 1} = 3,003$$

27. **(a)** Six applicants are selected from among 12 without regard to order. $C_{12,6} = \dfrac{12!}{6!6!} = \dfrac{479,001,600}{(720)^2} = 924$.

   (b) This problem is asking, "In how many ways can six women be selected from seven applicants?"

   $$C_{7,6} = \frac{7!}{6! \times 1!} = 7$$

   (c) $P(\text{event A}) = \dfrac{\text{number of favorable outcomes}}{\text{total number of outcomes}}$

   $P(\text{all hired are women}) = \dfrac{7}{924} = \dfrac{1}{132} \approx 0.008$

Chapter Review Problems

1. **(a)** The individual does not own a cell phone.
   (b) The individual owns both a cell phone and a laptop computer.
   (c) The individual owns either a cell phone or a laptop computer or both.
   (d) A laptop owner who owns a cell phone.
   (e) A cell phone owner who owns a laptop.

3. For independent events, $P(A \mid B) = P(A)$.

5. **(a)**
   $P(\text{Get Both Offers}) = P(\text{Get Offer on 1st and Get Offer on 2nd}) =$
   $P(\text{Get Offer on 1st})P(\text{Get Offer on 2nd}) = 0.7(0.8) = 0.56$

   So the chance of getting both offers is lower than getting either offer individually.

   (b)
   $P(\text{Get Either Offer}) = P(\text{Get Offer on 1st or Get Offer on 2nd}) =$
   $P(\text{Get Offer on 1st}) + P(\text{Get Offer on 2nd}) - P(\text{Get Both Offers}) = 0.7 + 0.8 - 0.56 = 0.94$

   The chance of getting either job is quite high, so it seems worthwhile to apply to both jobs.

7. **(a)** No, unless events A and B are independent. If they are not, we need either $P(A \mid B)$ or $P(B \mid A)$ to compute $P(A \text{ and } B)$.
   (b) Yes, now we can compute $P(A \text{ and } B) = P(A) \times P(B)$.

9. $P(\text{asked}) = 24\% = 0.24$
   $P(\text{received} \mid \text{asked}) = 45\% = 0.45$
   $P(\text{asked and received}) = P(\text{asked}) \times P(\text{received} \mid \text{asked}) = (0.24) \times (0.45) = 10.8\% = 0.108$

11. **(a)** Throw a large number of similar thumbtacks or one thumbtack a large number of times, and record the relative frequency of the outcomes. Assume that the thumbtack falls either flat side down or tilted. To estimate the probability the tack lands on its flat side, find the relative frequency of this occurrence, dividing the number of times this occurred by the total number of thumbtack tosses.

(b) The sample space consists of two outcomes: flat side down and tilted.

(c) $P(\text{flat side down}) = \dfrac{340}{500} = 0.68$

$P(\text{tilted}) = 1 - 0.68 = 0.32$

13. (a) Possible values for x are 2, 3, 4, 5, 6, 7, 8, 9, 10, 11, and 12.

(b) Below, the values for x are listed, along with the combinations required.

2	1 and 1	1 way
3	1 and 2, or 2 and 1	2 ways
4	1 and 3, 2 and 2, 3 and 1	3 ways
5	1 and 4, 2 and 3, 3 and 2, 4 and 1	4 ways
6	1 and 5, 2 and 4, 3 and 3, 4 and 2, 5 and 1	5 ways
7	1 and 6, 2 and 5, 3 and 4, 4 and 3, 5 and 2, 6 and 1	6 ways
8	2 and 6, 3 and 5, 4 and 4, 5 and 3, 6 and 2	5 ways
9	3 and 6, 4 and 5, 5 and 4, 6 and 3	4 ways
10	4 and 6, 5 and 5, 6 and 4	3 ways
11	5 and 6, 6 and 5	2 ways
12	6 and 6	1 way

x	$P(x)$
2	$\dfrac{1}{36} \approx 0.028$
3	$\dfrac{2}{36} \approx 0.056$
4	$\dfrac{3}{36} \approx 0.083$
5	$\dfrac{4}{36} \approx 0.111$
6	$\dfrac{5}{36} \approx 0.139$
7	$\dfrac{6}{36} \approx 0.167$
8	$\dfrac{5}{36} \approx 0.139$
9	$\dfrac{4}{36} \approx 0.111$
10	$\dfrac{3}{36} \approx 0.083$
11	$\dfrac{2}{36} \approx 0.056$
12	$\dfrac{1}{36} \approx 0.028$

Where there are $(6)(6) = 36$ possible, equally likely outcomes. (The sums, however, are not equally likely).

15. $C_{8,2} = \dfrac{8!}{2!6!} = \dfrac{8 \cdot 7 \cdot 6!}{(2 \cdot 1)6!} = \dfrac{56}{2} = 28$

17. Five multiple choice questions, each with four possible responses (A, B, C, or D).

There are $4 \times 4 \times 4 \times 4 \times 4 = 1{,}024$ possible sequences, such as A, D, B, B.

$P(\text{getting the correct sequence}) = \dfrac{1}{1024} \approx 0.00098$

19. There are 10 possible numbers per turn of dial and, we turn the dial three times.
There are $10 \times 10 \times 10 = 1{,}000$ possible combinations.

Chapter 5: The Binomial Probability Distribution and Related Topics

Section 5.1

1. **(a)** The number of traffic fatalities can be only a whole number. This is a discrete random variable.
   (b) Distance can assume any value, so this is a continuous random variable.
   (c) Time can take on any value, so this is a continuous random variable.
   (d) The number of ships can be only a whole number. This is a discrete random variable.
   (e) Weight can assume any value, so this is a continuous random variable.

3. **(a)** $\sum P(x) = 0.25 + 0.60 + 0.15 = 1.00$

   Yes, this is a valid probability distribution because the sum of the probabilities is 1, each probability is between 0 and 1 inclusive, and each event is assigned a probability.

   (b) $\sum P(x) = 0.25 + 0.60 + 0.20 = 1.05$

   No, this is not a probability distribution because the probabilities sum to more than 1.

5. No. Even though the outcomes in the sample space are the same, the individual probabilities may differ in a way that produces the same μ but a different standard deviation.

7. $\mu = 0(0.25) + 1(0.60) + 2(0.15) = 0.9$

   $$\sigma = \sqrt{\sum (x - \mu)^2 P(x)} = \sqrt{(0 - 0.9)^2 (0.25) + (1 - 0.9)^2 (0.60) + (2 - 0.9)^2 (0.15)} \approx 0.6245$$

9. **(a)** Yes, seven of the ten digits are assigned to "make a basket."
   (b) Let S represent "make a basket" and F represent "miss." We have

   $F\,F\,S\,S\,S\,F\,F\,F\,S\,S$

   (c) Yes, again, seven of the ten digits represent "make a basket." We have

   $S\,S\,S\,S\,S\,S\,S\,S\,S\,S$

11. **(a)** $\sum P(x) = 0.21 + 0.14 + 0.22 + 0.15 + 0.20 + 0.08 = 1.00$

    Yes, this is a valid probability distribution because the events are distinct and the probabilities sum to 1.

(b)

Histogram of Income Distribution

(c) $\mu = \sum xP(x)$

$= 10(0.21) + 20(0.14) + 30(0.22) + 40(0.15) + 50(0.20) + 60(0.08)$

$= 32.3$

(d) $\sigma = \sqrt{\sum (x-\mu)^2 P(x)}$

$= \sqrt{(-22.3)^2(0.21) + (-12.3)^2(0.14) + (-2.3)^2(0.22) + (7.7)^2(0.15) + (17.7)^2(0.20) + (27.7)^2(0.08)}$

$= \sqrt{259.71}$

≈ 16.12

13. (a)

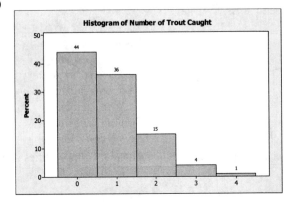

Histogram of Number of Trout Caught

(b) $P(1 \text{ or more}) = 1 - P(0)$

$= 1 - 0.44$

$= 0.56$

(c) $P(2 \text{ or more}) = P(2) + P(3) + P(4 \text{ or more})$

$= 0.15 + 0.04 + 0.01$

$= 0.20$

(d) $\mu = \sum xP(x)$

$= 0(0.44) + 1(0.36) + 2(0.15) + 3(0.04) + 4(0.01)$

$= 0.82$

(e) $\sigma = \sqrt{\Sigma(x-\mu)^2 P(x)}$

$$= \sqrt{(-0.82)^2(0.44) + (0.18)^2(0.36) + (1.18)^2(0.15) + (2.18)^2(0.04) + (3.18)^2(0.01)}$$

$$= \sqrt{0.8076}$$

$$\approx 0.899$$

15. (a) $P(\text{win}) = \dfrac{15}{719} \approx 0.021$

$P(\text{not win}) = \dfrac{719-15}{719} = \dfrac{704}{719} \approx 0.979$

(b) Expected earnings $= (\text{value of dinner})(\text{probability of winning})$

$$= \$35\left(\frac{15}{719}\right)$$

$$\approx \$0.73$$

Lisa's expected earnings are $0.73.

Contribution $= \$15 - \$0.73 = \$14.27$

Lisa effectively contributed $14.27 to the hiking club.

17. (a) $P(60 \text{ years}) = 0.01191$

Expected loss $= \$50,000(0.01191) = \$595.50$

The expected loss for Big Rock Insurance is $595.50.

(b)

Probability	Expected Loss
$P(61) = 0.01292$	$\$50,000(0.01292) = \$646$
$P(62) = 0.01396$	$\$50,000(0.01396) = \$698$
$P(63) = 0.01503$	$\$50,000(0.01503) = \$751.50$
$P(64) = 0.01613$	$\$50,000(0.01613) = \$806.50$

Expected loss $= \$595.50 + \$646 + \$698 + \$751.50 + \$806.50$
$$= \$3,497.50$$

The total expected loss is $3,497.50.

(c) $\$3,497.50 + \$700 = \$4,197.50$
They should charge $4,197.50.

(d) $\$5,000 - \$3,497.50 = \$1,502.50$
They can expect to make $1,502.50.

19. (a) $W = x_1 - x_2;\ a = 1,\ b = -1$
$\mu_w = \mu_1 - \mu_2 = 115 - 100 = 15$
$\sigma_w^2 = 1^2\sigma_1^2 + (-1)^2\sigma_2^2 = 12^2 + 8^2 = 208$
$\sigma_w = \sqrt{\sigma_w^2} = \sqrt{208} \approx 14.4$

(b) $W = 0.5x_1 + 0.5x_2$; $a = 0.5$, $b = 0.5$

$\mu_w = 0.5\mu_1 + 0.5\mu_2 = 0.5(115) + 0.5(100) = 107.5$

$\sigma_w^2 = (0.5)^2 \sigma_1^2 + (0.5)^2 \sigma_2^2 = 0.25(12)^2 + 0.25(8)^2 = 52$

$\sigma_w = \sqrt{\sigma_w^2} = \sqrt{52} \approx 7.2$

(c) $L = 0.8x_1 - 2$; $a = -2$, $b = 0.8$

$\mu_L = -2 + 0.8\mu_1 = -2 + 0.8(115) = 90$

$\sigma_L^2 = (0.8)^2 \sigma_1^2 = 0.64(12)^2 = 92.16$

$\sigma_L = \sqrt{\sigma_L^2} = \sqrt{92.16} = 9.6$

(d) $L = 0.95x_2 - 5$; $a = -5$, $b = 0.95$

$\mu_L = -5 + 0.95\mu_2 = -5 + 0.95(100) = 90$

$\sigma_L^2 = (0.95)^2 \sigma_2^2 = 0.9025(8)^2 = 57.76$

$\sigma_L = \sqrt{\sigma_L^2} = \sqrt{57.76} = 7.6$

21. (a) $W = 0.5x_1 + 0.5x_2$; $a = 0.5$, $b = 0.5$

$\mu_w = 0.5\mu_1 + 0.5\mu_2 = 0.5(50.2) + 0.5(50.2) = 50.2$

$\sigma_w^2 = 0.5^2 \sigma_1^2 + 0.5^2 \sigma_2^2 = 0.5^2 (11.5)^2 + 0.5^2 (11.5)^2 = 66.125$

$\sigma_w = \sqrt{\sigma_w^2} = \sqrt{66.125} \approx 8.13$

(b) Single policy (x_1): $\mu_1 = 50.2$

Two policies (W): $\mu_w \approx 50.2$

The means are the same.

(c) Single policy (x_1): $\sigma_1 = 11.5$

Two policies (W): $\sigma_w \approx 8.13$

The standard deviation for the average of two policies is smaller.

(d) Yes, the risk decreases by a factor of $\dfrac{1}{\sqrt{n}}$ because $\sigma_w = \dfrac{1}{\sqrt{n}}\sigma$.

23. (a) Essay. Answers vary.

(b) Square the standard deviations and substitute the results into the formula to evaluate c_1, c_2, and c_3.

(c)

$\sigma^2{}_w = (0.64)^2 (7)^2 + (0.22)^2 (12)^2 + (0.14)^2 (15)^2 = 31.45$

$\sigma_w = \sqrt{\sigma^2{}_w} = \sqrt{31.45} \approx 5.608$

Section 5.2

1. The random variable counts the number of successes that occur in the n trials.

3. Binomial experiments have two possible outcomes, denoted success and failure.

5. Any monitor failure might endanger patient safety, so you should be concerned about the probability of at least one failure, not just exactly one failure.

7. **(a)** No, there must be only two outcomes for each trial. Here, there are three outcomes.
  (b) Yes. If we combined outcomes B and C into a single outcome, then we have a binomial experiment. The probability of success for each trial is $P(A) = p = 0.40$.

9. **(a)** A trial is the random selection of one student and noting whether the student is a freshman or is not a freshman. Here, the probability of success is $p = 0.40$ and the probability of a failure is $1 - 0.40 = 0.60$
  (b) For a small population of size 30, sampling without replacement will alter the probability of drawing a freshman. In this situation, the hypergeometric distribution is appropriate.

11. $q = 1 - 0.30 = 0.70$
  (a) $P(r = 0) = C_{7,0} (0.30)^0 (0.70)^{7-0} \approx 0.082$
  (b) $P(r \geq 1) = 1 - P(r = 0) = 1 - 0.082 = 0.918$

13. $q = 1 - 0.85 = 0.15$
  (a) $P(r \leq 1) = P(r = 0) + P(r = 1) = C_{6,0} (0.85)^0 (0.15)^{6-0} + C_{6,1} (0.85)^1 (0.15)^{6-1} \approx 0.000399$
  (b) Yes, this probability is quite small. It would be a rare event to get only 1 or 2 successes when the probability of success on a single trial is so high.

15. A trial is one flip of a fair quarter. Success = head. Failure = tail.
  $n = 3$, $p = 0.5$, $q = 1 - 0.5 = 0.5$
  (a) $P(3) = C_{3,3} (0.5)^3 (0.5)^{3-3}$
  $$= 1(0.5)^3 (0.5)^0$$
  $$= 0.125$$
  To find this value in Table 3 of Appendix II, use the group in which $n = 3$, the column headed by $p = 0.5$ and the row headed by $r = 3$.
  (b) $P(2) = C_{3,2} (0.5)^2 (0.5)^{3-2}$
  $$= 3(0.5)^2 (0.5)^1$$
  $$= 0.375$$
  To find this value in Table 3 of Appendix II, use the group in which $n = 3$, the column headed by $p = 0.5$ and the row headed by $r = 2$.

  (c) $P(r \geq 2) = P(2) + P(3)$
  $$= 0.125 + 0.375$$
  $$= 0.5$$

  (d) The probability of getting exactly three tails is the same as getting exactly zero heads.
  $$P(0) = C_{3,0} (0.5)^0 (0.5)^{3-0}$$
  $$= 1(0.5)^0 (0.5)^3$$
  $$= 0.125$$
  To find this value in Table 3 of Appendix II, use the group in which $n = 3$, the column headed by $p = 0.5$ and the row headed by $r = 0$.

17. A trial consists of determining the sex of a wolf. Success = male. Failure = female.
    (a) $n = 12, p = 0.55, q = 0.45$

$$P(r \geq 6) = P(6) + P(7) + P(8) + P(9) + P(10) + P(11) + P(12)$$
$$= 0.212 + 0.223 + 0.170 + 0.092 + 0.034 + 0.008 + 0.001$$
$$= 0.740$$

Six or more females is the same as six or fewer males.
$$P(r \leq 6) = P(0) + P(1) + P(2) + P(3) + P(4) + P(5) + P(6)$$
$$= 0.000 + 0.001 + 0.007 + 0.028 + 0.076 + 0.149 + 0.212$$
$$= 0.473$$

Fewer than four females is the same as more than eight males.
$$P(r > 8) = P(9) + P(10) + P(11) + P(12)$$
$$= 0.092 + 0.034 + 0.008 + 0.001$$
$$= 0.135$$

    (b) $n = 12, p = 0.70, q = 0.30$
$$P(r \geq 6) = P(6) + P(7) + P(8) + P(9) + P(10) + P(11) + P(12)$$
$$= 0.079 + 0.158 + 0.231 + 0.240 + 0.168 + 0.071 + 0.014$$
$$= 0.961$$
$$P(r \leq 6) = P(0) + P(1) + P(2) + P(3) + P(4) + P(5) + P(6)$$
$$= 0.000 + 0.000 + 0.000 + 0.001 + 0.008 + 0.029 + 0.079$$
$$= 0.117$$
$$P(r > 8) = P(9) + P(10) + P(11) + P(12)$$
$$= 0.240 + 0.168 + 0.071 + 0.014$$
$$= 0.493$$

19. A trial consists of a woman's response regarding her mother-in-law. Success = dislike. Failure = like.
    $n = 6, p = 0.90, q = 1 - 0.90 = 0.10$

    (a) $P(6) = C_{6,6}(0.90)^6 (0.10)^{6-6}$
$$= 1(0.90)^6 (0.10)^0$$
$$= 0.531$$

    (b) $P(0) = C_{6,0}(0.90)^0 (0.10)^{6-0}$
$$= 1(0.90)^0 (0.10)^6$$
$$\approx 0.000 \ \text{(to 3 digits)}$$

    (c) $P(r \geq 4) = P(4) + P(5) + P(6)$
$$= 0.098 + 0.354 + 0.531$$
$$= 0.983$$

(d) $P(r \le 3) = 1 - P(r \ge 4)$

$\approx 1 - 0.983$

$= 0.017$

From the table:

$P(r \le 3) = P(0) + P(1) + P(2) + P(3)$

$= 0.000 + 0.000 + 0.001 + 0.015$

$= 0.016$

21. A trial consists of taking a polygraph examination. Success = pass. Failure = fail.
$n = 9$, $p = 0.85$, $q = 1 - 0.85 = 0.15$

(a) $P(9) = 0.232$

(b) $P(r \ge 5) = P(5) + P(6) + P(7) + P(8) + P(9)$

$= 0.028 + 0.107 + 0.260 + 0.368 + 0.232$

$= 0.995$

(c) $P(r \le 4) = 1 - P(r \ge 5)$

$= 1 - 0.995$

$= 0.005$

From the table:

$P(r \le 4) = P(0) + P(1) + P(2) + P(3) + P(4)$

$= 0.000 + 0.000 + 0.000 + 0.001 + 0.005$

$= 0.006$

The two results should be equal, but because of rounding error, they differ slightly.

(d)   All students failing is the same as no students passing.

$P(0) = 0.000$ (to 3 digits)

23. (a) A trial consists of using the Meyers-Briggs instrument to determine if a person in marketing is an extrovert. Success = extrovert. Failure = not extrovert.
$n = 15$, $p = 0.75$, $q = 1 - 0.75 = 0.25$

$P(r \ge 10) = P(10) + P(11) + P(12) + P(13) + P(14) + P(15)$

$= 0.165 + 0.225 + 0.225 + 0.156 + 0.067 + 0.013$

$= 0.851$

$P(r \ge 5) = P(5) + P(6) + P(7) + P(8) + P(9) + P(r \ge 10)$

$= 0.001 + 0.003 + 0.013 + 0.039 + 0.092 + 0.851$

$= 0.999$

$P(15) = 0.013$

(b) A trial consists of using the Meyers-Briggs instrument to determine if a computer programmer is an introvert. Success = introvert. Failure = not introvert.

$n = 5$, $p = 0.60$, $q = 1 - 0.60 = 0.40$

$P(0) = 0.010$

$$P(r \geq 3) = P(3) + P(4) + P(5)$$
$$= 0.346 + 0.259 + 0.078$$
$$= 0.683$$

25. A trial consists of the response from adults regarding their concern that Social Security numbers are used for general identification. Success = concerned that SS numbers are being used for identification. Failure = not concerned that SS numbers are being used for identification.

$n = 8$, $p = 0.53$, $q = 1 - 0.53 = 0.47$

(a) $P(r \leq 5) = P(0) + P(1) + P(2) + P(3) + P(4) + P(5)$
$$= 0.002381 + 0.021481 + 0.084781 + 0.191208 + 0.269521 + 0.243143$$
$$= 0.812515$$
$P(r \leq 5) = 0.81251$ from the cumulative probability is the same, truncated to 5 digits.

(b) $P(r > 5) = P(6) + P(7) + P(8)$
$$= 0.137091 + 0.044169 + 0.006726$$
$$= 0.187486$$
$P(r > 5) = 1 - P(r \leq 5)$
$$= 1 - 0.81251$$
$$= 0.18749$$
Yes, this is the same result rounded to 5 digits.

27. **(a)** $p = 0.30$, $P(3) = 0.132$
$p = 0.70$, $P(2) = 0.132$
They are the same.

(b) $p = 0.30$, $P(r \geq 3) = 0.132 + 0.028 + 0.002 = 0.162$
$p = 0.70$, $P(r \leq 2) = 0.002 + 0.028 + 0.132 = 0.162$
They are the same.

(c) $p = 0.30$, $P(4) = 0.028$
$p = 0.70$, $P(1) = 0.028$
$r = 1$

(d) The column headed by $p = 0.80$ is symmetric with the one headed by $p = 0.20$.

29. **(a)** $n = 8; p = 0.65$

$$P(6 \le r, \ given \ 4 \le r) = \frac{P(6 \le r \ and \ 4 \le r)}{P(4 \le r)}$$

$$= \frac{P(6 \le r)}{P(4 \le r)}$$

$$= \frac{P(6) + P(7) + P(8)}{P(4) + P(5) + P(6) + P(7) + P(8)}$$

$$= \frac{0.259 + 0.137 + 0.032}{0.188 + 0.279 + 0.259 + 0.137 + 0.032}$$

$$= \frac{0.428}{0.895}$$

$$= 0.478$$

(b) $n = 10; p = 0.65$

$$P(8 \le r, \ given \ 6 \le r) = \frac{P(8 \le r \ and \ 6 \le r)}{P(6 \le r)}$$

$$= \frac{P(8 \le r)}{P(6 \le r)}$$

$$= \frac{P(8) + P(9) + P(10)}{P(6) + P(7) + P(8) + P(9) + P(10)}$$

$$= \frac{0.176 + 0.072 + 0.014}{0.238 + 0.252 + 0.176 + 0.072 + 0.014}$$

$$= \frac{0.262}{0.752}$$

$$= 0.348$$

(c) Answers vary. Possibilities include the stock market, getting raises at work, or passing exams.

(d) Let event $A = 6 \le r$ and event $B = 4 \le r$ in the formula.

31. **(a)** $P(3,2,1) = \frac{6!}{3!2!1!}(0.5)^3(0.3)^2(0.2)^1 = 0.135$

(b) $P(4,2,0) = \frac{6!}{4!2!0!}(0.5)^4(0.3)^2(0.2)^0 \approx 0.084$

Section 5.3

1. The average number of successes in n trials.

3. **(a)** $\mu = np = 8(0.2) = 1.6$

$\sigma = \sqrt{npq} = \sqrt{8(0.2)(0.8)} \approx 1.13$

(b) $P(r \ge 5) = 0.010$ according to the table. Yes, this is unusual, as 5 successes is more than 2.5 standard deviations above the expected value.

5. **(a)**   Yes, 120 is more than 2.5 standard deviations above the mean.
 (b)   Yes, 40 is more than 2.5 standard deviations below the mean.
 (c)   No, the entire interval is within 2.5 standard deviations above and below the mean.

7. **(a)**   The distribution is symmetric.

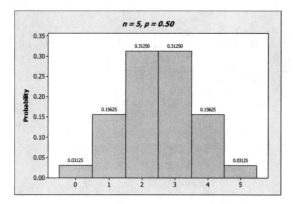

 (b)   The distribution is skewed right.

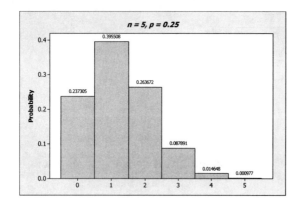

 (c)   The distribution is skewed left.

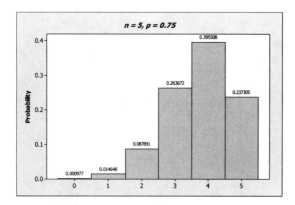

 (d)   The distributions are mirror images of one another.
 (e)   The distribution would be skewed left for $p = 0.73$ because $p > 0.50$.

9.   **(a)**   Skewed left

      (b) $\mu = np = 10(0.85) = 8.5$

      (c)   Very low. The expected number of successes in 10 trials is 8.5, and p is so high, that it would be unusual to have so few successes in 10 trials.

      (d)   Very high. The expected number of successes in 10 trials is 8.5, and p is so high, that it would be common to have 8 or more successes in 10 trials.

11.   **(a)** $n = 10$, $p = 0.80$

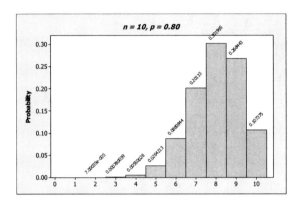

          $\mu = np = 10(0.8) = 8$

          $\sigma = \sqrt{npq} = \sqrt{10(0.8)(0.2)} \approx 1.26$

      (b) $n = 10$, $p = 0.5$

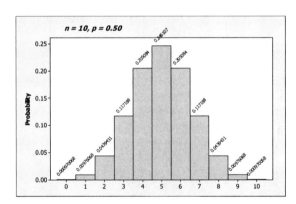

          $\mu = np = 10(0.5) = 5$

          $\sigma = \sqrt{npq} = \sqrt{10(0.5)(0.5)} \approx 1.58$

      (c)   Yes; since the graph in part (a) is skewed left, it supports the claim that more households buy film that have children under 2 years of age than households that have no children under 21 years of age.

13. (a) $n = 6$, $p = 0.70$

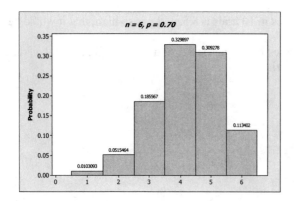

(b) $\mu = np = 6(0.70) = 4.2$

$\sigma = \sqrt{npq} = \sqrt{6(0.70)(0.30)} \approx 1.122$

We expect 4.2 friends to be found.

(c) Find n such that $P(r \geq 2) = 0.97$.

Try $n = 5$.

$$P(r \geq 2) = P(2) + P(3) + P(4) + P(5)$$
$$= 0.132 + 0.309 + 0.360 + 0.168$$
$$= 0.969$$
$$\approx 0.97$$

You would have to submit five names to be 97% sure that at least two addresses will be found.

If you solve this problem as

$$P(r \geq 2) = 1 - P(r < 2)$$
$$= 1 - \left[P(r = 0) + P(r = 1) \right]$$
$$= 1 - 0.002 - 0.028 = 0.97$$

the answers differ owing to rounding error in the table.

15. (a) $n = 7$, $p = 0.20$

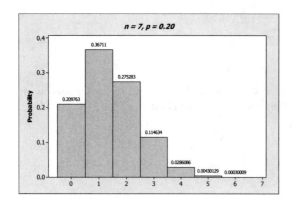

(b) $\mu = np = 7(0.20) = 1.4$

$\sigma = \sqrt{npq} = \sqrt{7(0.20)(0.80)} \approx 1.058$

We expect 1.4 people to be illiterate.

(c) Let success = literate and $p = 0.80$.
Find n such that $P(r \geq 7) = 0.98$.
Try $n = 12$.

$P(r \geq 7) = P(7) + P(8) + P(9) + P(10) + P(11) + P(12)$

$\qquad = 0.053 + 0.133 + 0.236 + 0.283 + 0.206 + 0.069$

$\qquad = 0.98$

You would need to interview 12 people to be 98% sure that at least 7 of these people are not illiterate.

17. (a) $n = 8,\ p = 0.25$

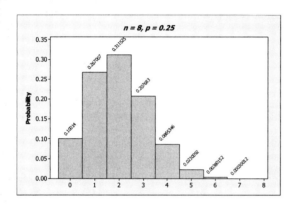

(b) $\mu = np = 8(0.25) = 2$

$\sigma = \sqrt{npq} = \sqrt{8(0.25)(0.75)} \approx 1.225$

We expect two people to believe that the product is actually improved.

(c) Find n such that $P(r \geq 1) = 0.99$.
Try $n = 16$.

$P(r \geq 1) = 1 - P(0)$

$\qquad = 1 - 0.01$

$\qquad = 0.99$

Sixteen people are needed in the marketing study to be 99% sure that at least one person believes the product to be improved.

19. (a) Since success = not a repeat offender, then $p = 0.75$.

r	0	1	2	3	4
$P(r)$	0.004	0.047	0.211	0.422	0.316

(b)

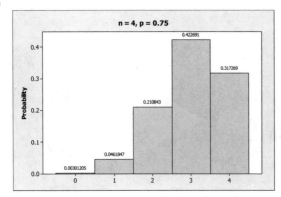

(c) $\mu = np = 4(0.75) = 3$

We expect three parolees to not repeat offend.

$\sigma = \sqrt{npq} = \sqrt{4(0.75)(0.25)} \approx 0.866$

(d)   Find n such that

$P(r \geq 3) = 0.98$

Try $n = 7$.

$P(r \geq 3) = P(3) + P(4) + P(5) + P(6) + P(7)$

$= 0.058 + 0.173 + 0.311 + 0.311 + 0.133$

$= 0.986$

This is slightly higher than needed, but $n = 6$ yields $P(r \geq 3) = 0.963$.

Alice should have a group of seven to be about 98% sure that three or more will not become repeat offenders.

21. **(a)**   Let success = available, then $p = 0.75$, $n = 12$. $P(12) = 0.032$.

(b)   Let success = not available, then $p = 0.25$, $n = 12$.

$P(r \geq 6) = P(6) + P(7) + P(8) + P(9) + P(10) + P(11) + P(12)$

$= 0.040 + 0.011 + 0.002 + 0.000 + 0.000 + 0.000 + 0.000$

$= 0.053$

(c) $n = 12, p = 0.75$

$\mu = np = 12(0.75) = 9$

The expected number of those available to serve on the jury is nine.

$\sigma = \sqrt{npq} = \sqrt{12(0.75)(0.25)} = 1.5$

(d) $p = 0.75$

Find n such that $P(r \geq 12) = 0.959$.

Try $n = 20$.

$P(r \geq 12) = P(12) + P(13) + P(14) + P(15) + P(16) + P(17) + P(18) + P(19) + P(20)$

$= 0.061 + 0.112 + 0.169 + 0.202 + 0.190 + 0.134 + 0.067 + 0.021 + 0.003$

$= 0.959$

The jury commissioner must contact 20 people to be 95.9% sure of finding at least 12 people who are available to serve.

23. Let success = case solved, then $p = 0.2$, $n = 6$.

    (a) $P(0) = 0.262$

    (b) $P(r \geq 1) = 1 - P(0)$
$$= 1 - 0.262$$
$$= 0.738$$

    (c) $\mu = np = 6(0.20) = 1.2$

        The expected number of crimes that will be solved is 1.2.

        $\sigma = \sqrt{npq} = \sqrt{6(0.20)(0.80)} \approx 0.98$ **(d)** Find n such that $P(r \geq 1) = 0.90$.   Try $n = 11$.

    $P(r \geq 1) = 1 - P(0)$
$$= 1 - 0.086$$
$$= 0.914$$

    $[Note: $ For $n = 10$, $P(r \geq 1) = 0.893.]$

    The police must investigate 11 property crimes before they can be at least 90% sure of solving one or more cases.

25. **(a)** Japan: $n = 7$, $p = 0.95$. $P(7) = 0.698$

        United States: $n = 7$, $p = 0.60$. $P(7) = 0.028$

    (b) Japan: $n = 7$, $p = 0.95$.

        $\mu = np = 7(0.95) = 6.65$

        $\sigma = \sqrt{npq} = \sqrt{7(0.95)(0.05)} \approx 0.58$

        United States: $n = 7$, $p = 0.60$.

        $\mu = np = 7(0.60) = 4.2$

        $\sigma = \sqrt{npq} = \sqrt{7(0.60)(0.40)} \approx 1.30$

        The expected number of guilty verdicts in Japan is 6.65, and in the United States it is 4.2.

    (c) United States: $p = 0.60$.

        Find n such that $P(r \geq 2) = 0.99$. Try $n = 8$.

        $P(r \geq 2) = 1 - \left[P(0) + P(1) \right]$
$$= 1 - (0.001 + 0.008)$$
$$= 0.991$$

        Japan: $p = 0.95$.

        Find n such that $P(r \geq 2) = 0.99$. Try $n = 3$.

        $P(r \geq 2) = P(2) + P(3)$
$$= 0.135 + 0.857$$
$$= 0.992$$

        Cover eight trials in the United States and three trials in Japan.

27. **(a)** $p = 0.40$

    Find n such that $P(r \geq 1) = 0.99$.   Try $n = 9$.

    $$P(r \geq 1) = 1 - P(0)$$
    $$= 1 - 0.010$$
    $$= 0.990$$

    The owner must answer nine inquiries to be 99% sure of renting at least one room.

    (b) $n = 25, p = 0.40$

    $\mu = np = 25(0.40) = 10$

    The expected number is 10 room rentals.

Section 5.4

1. The geometric distribution

3. No, since the approximation requires $n \geq 100$.

5. $P(3) = 0.40(0.60)^{3-1} = 0.144$

7. $\mu = np = 200(0.04) = 8$

    $$P(8) = \frac{e^{-8}8^8}{8!} \approx 0.1396$$

9. **(a)** Geometric probability distribution, $p = 0.77$.

    $$P(n) = p(1-p)^{n-1}$$
    $$P(n) = (0.77)(0.23)^{n-1}$$

    (b) $P(1) = (0.77)(0.23)^{1-1} = (0.77)(0.23)^0 = 0.77$

    (c) $P(2) = (0.77)(0.23)^{2-1} = (0.77)(0.23)^1 = 0.1771$

    (d) $P(3 \text{ or more tries}) = 1 - P(1) - P(2) = 1 - 0.77 - 0.1771 = 0.0529$

    (e) $\mu = \dfrac{1}{p} = \dfrac{1}{0.77} \approx 1.29$

    The expected number is 1.29, or 1, attempt to pass.

11. **(a)** Geometric probability distribution, $p = 0.80$.

    $$P(n) = p(1-p)^{n-1}$$
    $$P(n) = (0.80)(0.20)^{n-1}$$

    (b) $P(1) = (0.80)(0.20)^{1-1} = 0.80$

    $P(2) = (0.80)(0.20)^{2-1} = 0.16$

    $P(3) = (0.80)(0.20)^{3-1} = 0.032$

    (c) $P(n \geq 4) = 1 - P(1) - P(2) - P(3) = 1 - 0.80 - 0.16 - 0.032 = 0.008$

(d) $P(n) = (0.04)(0.96)^{n-1}$

$P(1) = (0.04)(0.96)^{1-1} = 0.04$

$P(2) = (0.04)(0.96)^{2-1} = 0.0384$

$P(3) = (0.04)(0.96)^{3-1} = 0.0369$

$P(n \geq 4) = 1 - P(1) - P(2) - P(3) = 1 - 0.04 - 0.0384 - 0.0369 = 0.8847$

13. (a) Geometric probability distribution, $p = 0.30$.

$P(n) = p(1-p)^{n-1}$

$P(n) = (0.30)(0.70)^{n-1}$

(b) $P(3) = (0.30)(0.70)^{3-1} = 0.147$

(c) $P(n > 3) = 1 - P(1) - P(2) - P(3) = 1 - 0.30 - (0.30)(0.70) - 0.147 = 1 - 0.30 - 0.21 - 0.147 = 0.343$

(d) $\mu = \dfrac{1}{p} = \dfrac{1}{0.30} = 3.33$

The expected number is 3.33, or 3, trips.

15. (a) The Poisson distribution would be a good choice because frequency of initiating social grooming is a relatively rare occurrence. It is reasonable to assume that the events are independent and that the variable is the number of times that one otter initiates social grooming in a fixed time interval.

$\lambda = \dfrac{1.7}{10 \text{ min}} \cdot \dfrac{3}{3} = \dfrac{5.1}{30 \text{ min}};\ \lambda = 5.1 \text{ per } 30 \text{ min interval}$

$P(r) = \dfrac{e^{-\lambda}\lambda^r}{r!} = \dfrac{e^{-5.1}(5.1)^r}{r!}$

(b) $P(4) = \dfrac{e^{-5.1}(5.1)^4}{4!} \approx 0.1719$

$P(5) = \dfrac{e^{-5.1}(5.1)^5}{5!} \approx 0.1753$

$P(6) = \dfrac{e^{-5.1}(5.1)^6}{6!} \approx 0.1490$

(c) $P(r \geq 4) = 1 - P(0) - P(1) - P(2) - P(3) = 1 - 0.0061 - 0.0311 - 0.0793 - 0.1348 = 0.7487$

(d) $P(r < 4) = P(0) + P(1) + P(2) + P(3) = 0.0061 - 0.0311 - 0.0793 + 0.1348 = 0.2513$

or $\quad P(r < 4) = 1 - P(r \geq 4) = 1 - 0.7487 = 0.2513$

17. (a) The Poisson distribution would be a good choice because frequency of births is a relatively rare occurrence. It is reasonable to assume that the events are independent and that the variable is the number of births (or deaths) for a community of a given population size.

(b) For 1,000 people, $\lambda = 16$ births; $\lambda = 8$ deaths.

By Table 4 in Appendix II:

$P(10 \text{ births}) = 0.0341$
$P(10 \text{ deaths}) = 0.0993$
$P(16 \text{ births}) = 0.0992$
$P(16 \text{ deaths}) = 0.0045$

(c) For 1,500 people,

$\lambda = \dfrac{16}{1,000} \cdot \dfrac{1.5}{1.5} = \dfrac{24}{1,500}$; $\lambda = 24$ births per 1,500 people

$\lambda = \dfrac{8}{1000} \cdot \dfrac{1.5}{1.5} = \dfrac{12}{1500}$; $\lambda = 12$ deaths per 1500 people

By Table 4 in Appendix II or a calculator:

$P(10 \text{ births}) = 0.00066$
$P(10 \text{ deaths}) = 0.1048$
$P(16 \text{ births}) = 0.02186$
$P(16 \text{ deaths}) = 0.0543$

(d) For 750 people,

$\lambda = \dfrac{16}{1,000} \cdot \dfrac{0.75}{0.75} = \dfrac{12}{750}$; $\lambda = 12$ births per 750 people

$\lambda = \dfrac{8}{1,000} \cdot \dfrac{0.75}{0.75} = \dfrac{6}{750}$; $\lambda = 6$ deaths per 750 people

$P(10 \text{ births}) = 0.1048$
$P(10 \text{ deaths}) = 0.0413$
$P(16 \text{ births}) = 0.0543$
$P(16 \text{ deaths}) = 0.0003$

19. (a) The Poisson distribution would be a good choice because frequency of gale-force winds is a relatively rare occurrence. It is reasonable to assume that the events are independent and that the variable is the number of gale-force winds in a given time interval.

(b) $\lambda = \dfrac{1}{60 \text{ hours}} \cdot \dfrac{1.8}{1.8} = \dfrac{1.8}{108 \text{ hours}}$; $\lambda = 1.8$ per 108 hours

From Table 4 in Appendix II:

$P(2) = 0.2678$

$P(3) = 0.1607$

$P(4) = 0.0723$

$P(r < 2) = P(0) + P(1) = 0.1653 + 0.2975 = 0.4628$

(c) $\lambda = \dfrac{1}{60 \text{ hours}} \cdot \dfrac{3}{3} = \dfrac{3}{180 \text{ hours}}$; $\lambda = 3$ per 180 hours

$P(3) = 0.2240$
$P(4) = 0.1680$
$P(5) = 0.1008$
$P(r < 3) = P(0) + P(1) + P(2) = 0.0498 + 0.1494 + 0.2240 = 0.4232$

21. (a) The Poisson distribution would be a good choice because frequency of commercial building sales is a relatively rare occurrence. It is reasonable to assume that the events are independent and the variable is the number of buildings sold in a given time interval.

(b) $\lambda = \dfrac{8}{275 \text{ days}} \cdot \dfrac{\frac{12}{55}}{\frac{12}{55}} \approx \dfrac{\frac{96}{55}}{60 \text{ days}}$; $\lambda = \dfrac{96}{55} \approx 1.7$ per 60 days

From Table 4 in Appendix II:

$P(0) = 0.1827$

$P(1) = 0.3106$

$P(r \geq 2) = 1 - P(0) - P(1) = 1 - 0.1827 - 0.3106 = 0.5067$

(c) $\lambda = \dfrac{8}{275 \text{ days}} \cdot \dfrac{\frac{18}{55}}{\frac{18}{55}} \approx \dfrac{2.6}{90 \text{ days}}$; $\lambda \approx 2.6$ per 90 days

$P(0) = 0.0743$

$P(2) = 0.2510$

$P(r \geq 3) = 1 - P(0) - P(1) - P(2) = 1 - 0.0743 - 0.1931 - 0.2510 = 0.4816$

23. (a) The problem satisfies the conditions for a binomial experiment with

n large, $n = 1000$, and p small, $p = \frac{1}{569} \approx 0.0018$.
$np \approx 1000(0.0018) = 1.8 < 10$.

The Poisson distribution would be a good approximation to the binomial.

$\lambda = np \approx 1.8$

(b) From Table 4 in Appendix II, $P(0) = 0.1653$.

(c) $P(r > 1) = 1 - P(0) - P(1) = 1 - 0.1653 - 0.2975 = 0.5372$

(d) $P(r > 2) = P(r > 1) - P(2) = 0.5372 - 0.2678 = 0.2694$

(e) $P(r > 3) = P(r > 2) - P(3) = 0.2694 - 0.1607 = 0.1087$

25. (a) The problem satisfies the conditions for a binomial experiment with n large, $n = 175$, and p small, $p = 0.005$. $np = (175)(0.005) = 0.875 < 10$. The Poisson distribution would be a good approximation to the binomial. $n = 175$, $p = 0.005$, $\lambda = np = 0.9$.

(b) From Table 4 in Appendix II, $P(0) = 0.4066$.

(c) $P(r \geq 1) = 1 - P(0) = 1 - 0.4066 = 0.5934$

(d) $P(r \geq 2) = P(r \geq 1) - P(1) = 0.5934 - 0.3659 = 0.2275$

27. (a) $n = 100, p = 0.02, r = 2$

$$P(r) = C_{n,r}\, p^r (1-p)^{n-r}$$
$$P(2) = C_{100,2}(0.02)^2 (0.98)^{100-2} = 4950(0.0004)(0.1381) = 0.2734$$

(b) $\lambda = np = 100(0.02) = 2$

From Table 4 in Appendix II, $P(2) = 0.2707$.

(c) Yes, the approximation is correct to two decimal places.

(d) $n = 100;\ p = 0.02;\ r = 3$

By the formula for the binomial distribution,

$$P(3) = C_{100,3}(0.02)^3 (0.98)^{100-3} = 161{,}700(0.000008)(0.1409) = 0.1823$$

By the Poisson approximation, $\lambda = 3$, $P(3) = 0.1804$. This is correct to two decimal places.

29. (a) The Poisson distribution would be a good choice because hail storms in western Kansas are relatively rare occurrences. It is reasonable to assume that the events are independent and that the variable is the number of hailstorms in a fixed-square-mile area.

$$\lambda = \frac{2.1}{5} \cdot \frac{\frac{8}{5}}{\frac{8}{5}} = \frac{2.1\left(\frac{8}{5}\right)}{8} \approx \frac{3.4}{8}$$

$\lambda = 3.4$ storms per 8 square miles

(b) $P(r \geq 4,\ \text{given } r \geq 2) = \dfrac{P(r \geq 4 \text{ and } r \geq 2)}{P(r \geq 2)} = \dfrac{P(r \geq 4)}{P(r \geq 2)} = \dfrac{1 - P(r \leq 3)}{1 - P(r \leq 1)}$

$$= \frac{1 - (0.0334 + 0.1135 + 0.1929 + 0.2186)}{1 - (0.0334 + 0.1135)} = \frac{0.4416}{0.8531} = 0.5176$$

(c) $P(r < 6,\ \text{given } r \geq 3) = \dfrac{P(r < 6 \text{ and } r \geq 3)}{P(r \geq 3)} = \dfrac{P(r = 3) + P(r = 4) + P(r = 5)}{1 - P(r \leq 2)}$

$$= \frac{0.2186 + 0.1858 + 0.1264}{1 - (0.0334 + 0.1135 + 0.1929)} = \frac{0.5308}{0.6602} = 0.8040$$

31. (a) We have binomial trials for which the probability of success is $p = 0.80$ and failure is $q = 0.20$; $k = 12$ is a fixed whole number ≥ 1; n is a random variable representing the number of contacts needed to get the twelfth sale.

$$P(n) = C_{n-1,\, k-1}\, p^k q^{n-k}$$
$$P(n) = C_{n-1,\, 11}(0.80^{12})(0.20^{n-12})$$

(b) $P(12) = C_{11,11}(0.80^{12})(0.20^{0}) \approx 0.0687$

$P(13) = C_{12,11}(0.80^{12})(0.20^{1}) \approx 0.1649$

$P(14) = C_{13,11}(0.80^{12})(0.20^{2}) \approx 0.2144$

(c) $P(12 \leq n \leq 14) = P(12) + P(13) + P(14) = 0.0687 + 0.1649 + 0.2144 = 0.4480$

(d) $P(n > 14) = 1 - P(n \leq 14) = 1 - P(12 \leq n \leq 14) = 1 - 0.4480 = 0.5520$

(e) $\mu = \dfrac{k}{p} = \dfrac{12}{0.80} = 15$

$\sigma = \dfrac{\sqrt{kq}}{p} = \dfrac{\sqrt{12(0.20)}}{0.80} \approx 1.94$

The expected contact in which the twelfth sale will occur is the fifteenth contact, with a standard deviation of 1.94.

33. (a) This is binomial with $n-1$ trials and probability of success p. Thus we use the binomial probability distribution:

$P(A) = C_{n-1,k-1} * p^{k-1} q^{n-1-(k-1)}$

(b) $P(B) =$ success on one trial, namely, the nth trial $= p$ (by definition).

(c) By the definition of independent trials, $P(A \text{ and } B) = P(A) \times P(B)$.

(d) $P(A \text{ and } B) =$

$C_{n-1,k-1} \times p^{k-1} q^{n-1-(k-1)} \times p =$

$C_{n-1,k-1} \times p^{k} q^{n-k}$

(e) The results are the same.

Chapter Review Problems

1. A description of all the values of a random variable x, the associated probabilities for each value of x, the summation of the probabilities equal 1, and each probability takes on values between 0 and 1 inclusive.

3. (a) Yes, we expect $np = 10 \times 0.2 = 2$ successes. The standard deviation is $\sigma = 1.26$, so the boundary is $2 + (2.5) \times (1.26) = 5.15$. Thus six successes is unusual.

(b) No, $P(x > 5) = 0.0064$ (using a TI-83 calculator).

5. (a) $\mu = \sum x P(x) = 18.5(0.127) + 30.5(0.371) + 42.5(0.285) + 54.5(0.215) + 66.5(0.002) \approx 37.63$

The expected lease term is about 38 months.

$\sigma = \sqrt{\sum (x - \mu)^2 P(x)}$

$= \sqrt{(-19.13)^2 (0.127) + (-7.13)^2 (0.371) + (4.87)^2 (0.285) + (16.87)^2 (0.215) + (28.87)^2 (0.002)}$

$\approx \sqrt{134.95} \approx 11.6$ (using $\mu = 37.63$ in the calculations)

(b)

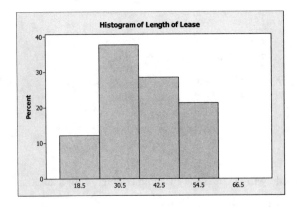

7.   This is a binomial experiment with 10 trials. A trial consists of a claim.
      Success = submitted by a male under 25 years of age.
      Failure = not submitted by a male under 25 years of age.

   (a)   The probabilities can be taken directly from Table 3 in Appendix II: $n = 10, p = 0.55$.

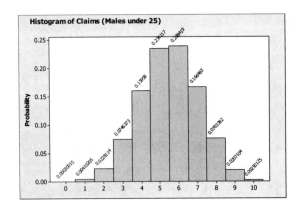

   (b) $P(x \geq 6) = P(6) + P(7) + P(8) + P(9) + P(10) = 0.504$
   (c) $\mu = np = 10(0.55) = 5.5$

          The expected number of claims made by males under age 25 is 5.5.

          $\sigma = \sqrt{npq} = \sqrt{10(0.55)(0.45)} \approx 1.57$

9. $n = 16, p = 0.50$
   (a) $P(r \geq 12) = P(12) + P(13) + P(14) + P(15) + P(16) = 0.028 + 0.009 + 0.002 + 0.000 + 0.000 = 0.039$

   (b) $P(r \leq 7) = P(0) + P(1) + P(2) + P(3) + P(4) + P(5) + P(6) + P(7)$
              $= 0.000 + 0.000 + 0.002 + 0.009 + 0.028 + 0.067 + 0.122 + 0.175 = 0.403$
   (c) $\mu = np = 16(0.50) = 8$

          The expected number of inmates serving time for dealing drugs is eight.

11. $n = 10, p = 0.75$

(a) The probabilities can be obtained directly from Table 3 in Appendix II.

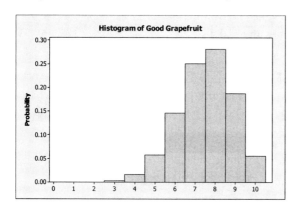

(b) No more than one is bad is the same event as at least nine are good.

$P(r \geq 9) = P(9) + P(10) = 0.188 + 0.056 = 0.244$

$P(r \geq 1) = P(1) + P(2) + P(3) + P(4) + P(5) + P(6) + P(7) + P(8) + P(9) + P(10)$

$\quad = 0.000 + 0.000 + 0.003 + 0.016 + 0.058 + 0.146 + 0.250 + 0.282 + 0.188 + 0.056 = 0.999$

(c) $\mu = np = 10(0.75) = 7.5$

We expect 7.5 good grapefruits.

(d) $\sigma = \sqrt{npq} = \sqrt{10(0.75)(0.25)} \approx 1.37$

13. $p = 0.85, n = 12$

$P(r \leq 2) = P(0) + P(1) + P(2) = 0.000 + 0.000 + 0.000 = 0.000$ (to 3 digits)

The data seem to indicate that the percent favoring the increase in fees is less than 85%.

15. (a) The Poisson distribution would be a good choice because coughs are a relatively rare occurrence. It is reasonable to assume that they are independent events, and the variable is the number of coughs in a fixed time interval.

(b) $\lambda = 11$ per 1 minute

From Table 4 in Appendix II,

$P(r \leq 3) = P(0) + P(1) + P(2) + P(3) = 0.0000 + 0.0002 + 0.0010 + 0.0037 = 0.0049$

(c) $\lambda = \dfrac{11}{60 \text{ seconds}} \cdot \dfrac{0.5}{0.5} = \dfrac{5.5}{30 \text{ seconds}}; \ \lambda = 5.5$ per 30 seconds

$P(r \geq 3) = 1 - P(0) - P(1) - P(2) = 1 - 0.0041 - 0.0225 - 0.0618 = 0.9116$

17. The loan-default problem satisfies the conditions for a binomial experiment. Moreover, p is small, n is large, and $np < 10$. Using the Poisson approximation to the binomial distribution is appropriate.

$n = 300, p = \dfrac{1}{350} = 0.0029, \lambda = np = 300(0.0029) \approx 0.86 \approx 0.9$

From Table 4 in Appendix II,

$P(r \geq 2) = 1 - P(0) - P(1) = 1 - 0.4066 - 0.3659 = 0.2275$

19. **(a)** Use the geometric distribution with $p = 0.5$.

$$P(n = 2) = (0.5)(0.5) = (0.5)^2 = 0.25$$

$P(n = 3) = (0.5)(0.5)(0.5) = 0.125$

$P(n = 4) = (0.5)(0.5)(0.5)(0.5) = 0.0625$

This is the geometric probability distribution with $p = 0.5$.

(b) $P(4) = (0.5)(0.5)^3 = (0.5)^4 = 0.0625$

$$P(n > 4) = 1 - P(1) - P(2) - P(3) - P(4) = 1 - 0.5 - 0.5^2 - 0.5^3 - 0.5^4 = 0.0625$$

Chapter 6: Normal Curves and Sampling Distributions

Section 6.1

1. **(a)** The curve is not normal; it is skewed left.
    (b) The curve is not normal; the right tail dips below the horizontal axis.
    (c) The curve is not normal; it is not bell-shaped, unimodal.
    (d) The curve is not normal; it is not smooth.

3. Figure 6-12 has the larger standard deviation. The mean of Figure 6-12 is 10, and the mean of Figure 6-13 is 4.

5. **(a)** 50%; the normal curve is symmetric about μ.
    (b) 68%
    (c) 99.7%

7. **(a)** $\mu = 65$, so 50% are taller than 65 inches.
    (b) $\mu = 65$, so 50% are shorter than 65 inches.
    (c) $\mu - \sigma = 65 - 2.5 = 62.5$ inches and $\mu + \sigma = 65 + 2.5 = 67.5$ inches, so 68% of college women are between 62.5 and 67.5 inches tall.
    (d) $\mu - 2\sigma = 65 - 2(2.5) = 65 - 5 = 60$ inches and $\mu - 2\sigma = 65 + 2(2.5) = 65 + 5 = 70$ inches, so 95% of college women are between 60 and 70 inches tall.

9. **(a)** $\mu - \sigma = 1{,}243 - 36 = 1{,}207$ and $\mu + \sigma = 1{,}243 + 36 = 1279$, so about 68% of the tree rings will date between 1207 and 1279 A.D.
    (b) $\mu - 2\sigma = 1243 - 2(36) = 1{,}171$ and $\mu + 2\sigma = 1243 + 2(36) = 1{,}315$, so about 95% of the tree rings will date between 1171 and 1315 A.D.
    (c) $\mu - 3\sigma = 1243 - 3(36) = 1{,}135$ and $\mu + 3\sigma = 1243 + 3(36) = 1{,}351$, so 99.7% (almost all) of the tree rings will date between 1135 and 1351 A.D.

11. **(a)** $\mu - \sigma = 3.15 - 1.45 = 1.70$ and $\mu + \sigma = 3.15 + 1.45 = 4.60$, so 68% of the experimental group will have a pain threshold between 1.70 and 4.60 mA.
    (b) $\mu - 2\sigma = 3.15 - 2(1.45) = 0.25$ and $\mu + 2\sigma = 3.15 + 2(1.45) = 6.05$, so 95% of the experimental group will have a pain threshold between 0.25 and 6.05 mA.

13. **(a)**

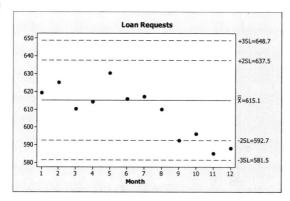

The economy would appear to be cooling off, as evidenced by an overall downward trend. Out-of-control warning signal III is present: Two of the last three consecutive points are below $\mu - 2\sigma = 592.7$.

(b)

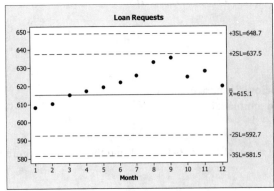

Here, it looks like the economy was heating up during months 1–9 and perhaps cooling off during months 10–12. Out-of-control warning signal II is present. There is a run of nine consecutive points above $\mu = 615.1$.

15.

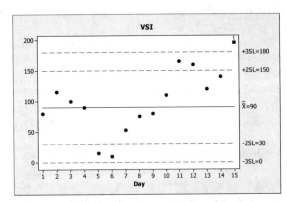

Out-of-control warning signals I and III are present. Day 15's VSI exceeds $\mu + 3\sigma$. Two of three consecutive points (days 10, 11, 12 or days 11, 12, 13) are about $\mu + 2\sigma = 150$, and two of three consecutive points (days 4, 5, 6 or days 5, 6, 7) are below $\mu - 2\sigma = 30$. Days 10–15 all show above-average air pollution levels; days 11, 12, and 15 triggered out-of-control signals, indicating that pollution abatement procedures should be in place.

17. (a) $P(-0.05 \le x < 0.03) = \dfrac{0.03 - (-0.05)}{0.05 - (-0.05)} = 0.80$

(b) $P(-0.02 < x \le 0.05) = \dfrac{0.05 - (-0.02)}{0.05 - (-0.05)} = 0.70$

(c) $P(-0.04 < x < 0.01) = \dfrac{0.01 - (-0.04)}{0.05 - (-0.05)} = 0.50$

(d) $\mu = \dfrac{\alpha + \beta}{2} = \dfrac{-0.05 + 0.05}{2} = 0$

$\sigma = \dfrac{\beta - \alpha}{\sqrt{12}} = \dfrac{0.05 - (-0.05)}{\sqrt{12}} \approx 0.0289$

If indeed the measurements have a mean of 0, then the measurements are unbiased. This must be based on actual measurements, not the assumption that measurements follow a uniform distribution from -0.05 to 0.05.

19. **(a)** $P(60 < x < \infty) = e^{-60/75} - e^{-\infty/75} \approx 0.4493$

(b) $P(0 < x < 140) = e^{-0/75} - e^{-140/75} \approx 0.8454$

(c) $P(60 < x < 100) = e^{-60/75} - e^{-100/75} \approx 0.1857$

(d) $P(0 < x < c) = 0.80$ is the equation to solve for c since that gives the boundaries under the exponential curve to give an 80% probability as required to meet town demand.

$$P(0 < x < c) = 0.80$$
$$e^{-0/75} - e^{-c/75} = 0.80$$
$$1 - e^{-c/75} = 0.80$$
$$0.20 = e^{-c/75}$$
$$\ln 0.20 = \ln\left(e^{-c/75}\right)$$
$$\ln 0.20 = -c/75$$
$$c = -75\ln 0.20 \approx 120.71$$

Section 6.2

1. The standard score is the number of standard deviations the measurement is from the mean.

3. The value is zero.

5. **(a)** $z = \dfrac{x - \mu}{\sigma} = \dfrac{25 - 30}{5} = -1$

(b) $z = \dfrac{x - \mu}{\sigma} = \dfrac{42 - 30}{5} = 2.4$

(c) $-2 = \dfrac{x - 30}{5}; \quad x = -2(5) + 30 = 20$

(d) $1.3 = \dfrac{x - 30}{5}; \quad x = 1.3(5) + 30 = 36.5$

7. The score of 40 is 1 standard deviation below its mean. The score of 45 is also 1 standard deviation below its mean. Thus they are the same in this regard.

9. **(a)** z scores > 0 indicate that the student scored above the mean: Robert, Juan, and Linda.
(b) z scores $= 0$ indicates that the student scored at the mean: Joel.
(c) z scores < 0 indicate that the student scored below the mean: Jan and Susan.

(d) $z = \dfrac{x - \mu}{\sigma}$, so $x = \mu + z\sigma$.

In this case, $x = 150 + z(20)$.

Robert $\quad x = 150 + 1.10(20) = 172$
Joel $\qquad x = 150 + 0(20) = 150$
Jan $\qquad x = 150 - 0.80(20) = 134$
Juan $\qquad x = 150 + 1.70(20) = 184$
Susan $\quad x = 150 - 2.00(20) = 110$
Linda $\quad x = 150 + 1.60(20) = 182$

11. $z = \dfrac{x - \mu}{\sigma}$, so in this case, $z = \dfrac{x - 4.8}{0.3}$.

(a) $4.5 < x$

$\dfrac{4.5 - 4.8}{0.3} < \dfrac{x - 4.8}{0.3}$ Subtract μ; divide by σ.

$-1.00 < z$

(b) $x < 4.2$

$\dfrac{x - 4.8}{0.3} < \dfrac{4.2 - 4.8}{0.3}$ Subtract μ; divide by σ.

$z < -2.00$

(c) $4.0 < x < 5.5$

$\dfrac{4.0 - 4.8}{0.3} < \dfrac{x - 4.8}{0.3} < \dfrac{5.5 - 4.8}{0.3}$ Subtract μ; divide by σ.

$-2.67 < z < 2.33$

For (d)–(f): Since $z = \dfrac{x - 4.8}{0.3}$, $x = 4.8 + 0.3z$.

(d) $z < -1.44$

$0.3z < 0.3(-1.44)$ Multiply by σ.

$4.8 + 0.3z < 4.8 + 0.3(-1.44)$ Add μ.

$x < 4.4$

(e) $1.28 < z$

$0.3(1.28) < 0.3z$ Multiply by σ.

$4.8 + 0.3(1.28) < 4.8 + 0.3z$ Add μ.

$5.2 < x$

(f) $-2.25 < z < -1.00$

$0.3(-2.25) < 0.3z < 0.3(-1.00)$ Multiply by σ.

$4.8 + 0.3(-2.25) < 4.8 + 0.3z < 4.8 + 0.3(-1.00)$ Add μ.

$4.1 < x < 4.5$

(g)   If the RBC were 5.9 or higher, that would be an unusually high red blood cell count.

$x \geq 5.9$

$\dfrac{x - 4.8}{0.3} \geq \dfrac{5.9 - 4.8}{0.3}$

$z \geq 3.67$ (a very large z value)

For Problems 13–49, refer to the following sketch patterns for guidance in calculations.

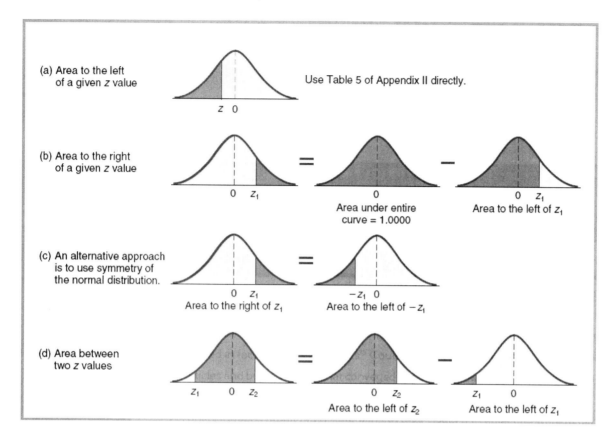

Use the left-tail style standard normal distribution table (see figures above).

(a) For areas to the *left* of a specified z value, use the table entry directly.

(b) For areas to the *right* of a specified z value, look up the table entry for z and subtract the table value from 1. *or* use the fact that the normal curve is symmetric about the mean, 0. The area in the right tail above a z value is the same as the area in the left tail below the value of $-z$. Thus, to find the area to the right of z, look up the table value for $-z$.

(c) For areas *between* two z values, z_1 and z_2, where $z_1 < z_2$, subtract the tabled value for z_1 from the tabled value for z_2.

These sketches and rules for finding the area for probability from the standard normal table apply for *any z*.

Student sketches should resemble those indicated with negative z values to left of 0 and positive z values to the right of zero.

13. Refer to figure (b).
The area to the right of $z = 0$ is 1 – area to left of $z = 0$, or $1 - 0.5000 = 0.5000$.

15. Refer to figure (a).
The area to the left of $z = -1.32$ is 0.0934.

17. Refer to figure (a).
The area to left of $z = 0.45$ is 0.6736.

19. Refer to figure (b).
The area to right of $z = 1.52$ is $1 - 0.9357 = 0.0643$.

21. Refer to figure (b).
The area to right of $z = -1.22$ is $1 - 0.1112 = 0.8888$.

23. Refer to figure (d).
The area between $z = 0$ and $z = 3.18$ is $0.9993 - 0.5000 = 0.4993$.

25. Refer to figure (d).
The area between $z = -2.18$ and $z = 1.34$ is $0.9099 - 0.0146 = 0.8953$.

27. Refer to figure (d).
The area between $z = 0.32$ and $z = 1.92$ is $0.9726 - 0.6255 = 0.3471$.

29. Refer to figure (d).
The area between $z = -2.42$ and $z = -1.77$ is $0.0384 - 0.0078 = 0.0306$.

31. Refer to figure (a).
$P(z \leq 0) = 0.5000$

33. Refer to figure (a).
$P(z \leq -0.13) = 0.4483$ (direct read)

35. Refer to figure (a).
$P(z \leq 1.20) = 0.8849$

37. Refer to figure (b).
$P(z \geq 1.35) = 1 - P(z < 1.35) = 1 - 0.9115 = 0.0885$

39. Refer to figure (b).
$P(x \geq -1.20) = 1 - P(z < -1.20) = 1 - 0.1151 = 0.8849$

41. Refer to figure (d).
$P(-1.20 \leq z \leq 2.64) = P(z \leq 2.64) - P(z < -1.20) = 0.9959 - 0.1151 = 0.8808$

43. Refer to figure (d).
$P(-2.18 \leq z \leq -0.42) = P(z \leq -0.42) - P(z < -2.18) = 0.3372 - 0.0146 = 0.3226$

45. Refer to figure (d).
$P(0 \leq z \leq 1.62) = P(z \leq 1.62) - P(z < 0) = 0.9474 - 0.5000 = 0.4474$

47. Refer to figure (d).
$P(-0.82 \leq z \leq 0) = P(z \leq 0) - P(z < -0.82) = 0.5000 - 0.2061 = 0.2939$

49. Refer to figure (d).
$P(-0.45 \leq z \leq 2.73) = P(z \leq 2.73) - P(z < -0.45) = 0.9968 - 0.3264 = 0.6704$

Section 6.3

1. Since 30 is the mean of this distribution, $P(x > 30) = 0.50$.

3. If 5% of the area lies to the left, we are in the lower tail. Thus z is negative.

5. We are given $\mu = 4$ and $\sigma = 2$. Since $z = \dfrac{x-\mu}{\sigma}$, we have $z = \dfrac{x-4}{2}$.

$P(3 \le x \le 6)$

$\quad = P(3 - 4 \le x - 4 \le 6 - 4)$ Subtract $\mu = 4$ from each part of the inequality.

$\quad = P\left(\dfrac{3-4}{2} \le \dfrac{x-4}{2} \le \dfrac{6-4}{2}\right)$ Divide each part by $\sigma = 2$.

$\quad = P\left(-\dfrac{1}{2} \le z \le \dfrac{2}{2}\right) = P(-0.5 \le z \le 1)$

$\quad = P(z \le 1) - P(z < -0.5)$ Refer to sketch (c) in the solutions for Section 6.2.

$\quad = 0.8413 - 0.3085 = 0.5328$

7. We are given $\mu = 40$ and $\sigma = 15$. Since $z = \dfrac{x-\mu}{\sigma}$, we have $z = \dfrac{x-40}{15}$.

$P(50 \le x \le 70)$

$\quad = P(50 - 40 \le x - 40 \le 70 - 40)$ Subtract $\mu = 40$.

$\quad = P\left(\dfrac{50-40}{15} \le \dfrac{x-40}{15} \le \dfrac{70-40}{15}\right)$ Divide by $\sigma = 15$.

$\quad = P(0.67 \le z \le 2) = P(z \le 2) - P(z < 0.67) = 0.9772 - 0.7486 = 0.2286$

9. We are given $\mu = 15$ and $\sigma = 3.2$. Since $z = \dfrac{x-\mu}{\sigma}$, we have $z = \dfrac{x-15}{3.2}$.

$P(8 \le x \le 12)$

$\quad = P(8 - 15 \le x - 15 \le 12 - 15)$ Subtract $\mu = 15$.

$\quad = P\left(\dfrac{8-15}{3.2} \le \dfrac{x-15}{3.2} \le \dfrac{12-15}{3.2}\right)$ Divide by $\sigma = 3.2$.

$\quad = P(-2.19 \le z \le -0.94) = P(z \le -0.94) - P(z < -2.19) = 0.1736 - 0.0143 = 0.1593$

11. We are given $\mu = 20$ and $\sigma = 3.4$. Since $z = \dfrac{x-\mu}{\sigma}$, we have $z = \dfrac{x-20}{3.4}$.

$P(x \ge 30)$

$\quad = P(x - 20 \ge 30 - 20)$ Subtract $\mu = 20$.

$\quad = P\left(\dfrac{x-20}{3.4} \ge \dfrac{30-20}{3.4}\right)$ Divide by $\sigma = 3.4$.

$\quad = P(z \ge 2.94) = 1 - P(z < 2.94) = 1 - 0.9984 = 0.0016$

13. We are given $\mu = 100$ and $\sigma = 15$. Since $z = \dfrac{x-\mu}{\sigma}$, we have $z = \dfrac{x-100}{15}$.

$P(x \ge 90)$

$\quad = P\left(\dfrac{x-100}{15} \ge \dfrac{90-100}{15}\right)$ Subtract μ; divide by σ.

$\quad = P(z \ge -0.67) = 1 - P(z < -0.67) = 1 - 0.2514 = 0.7486$

For Problems 15–23, refer to the following sketch patterns for guidance in calculation.

Inverse Normal: Use Table 5 of Appendix II to Find *z* Corresponding to a Given Area *A* (0 < *A* < 1)

(a) **Left-tail case:**
The given area *A*
is to the left of *z*.

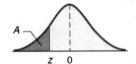 or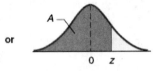

For the left-tail case, look up the
number *A* in the body of the table
and use the corresponding *z* value.

(b) **Right-tail case:**
The given area *A*
is to the right of *z*.

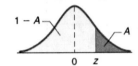 or

For the right-tail case, look up the
number 1 − *A* in the body of the table
and use the corresponding *z* value.

(c) **Center case:**
The given area *A* is
symmetric and centered
above *z* = 0. Half
of *A* lies to the left
and half lies to the
right of *z* = 0.

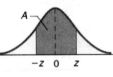

For the center case, look up the number $\frac{1-A}{2}$
in the body of the table and use the
corresponding ± *z* value.

Student sketches should resemble these figures, with negative *z* values to the left of zero and positive
z values to the right of zero and *A* written as a decimal.

15. Refer to figure (a).
Find *z* so that the area *A* to the left of *z* is 6% = 0.06. Since *A* = 0.06 is less than 0.5000, look for a negative *z*
value. *A* to left of −1.55 is 0.0606, and *A* to left of −1.56 is 0.0594. Since 0.06 is in the middle of 0.0606 and
0.0594, for our *z* value we will use the average of −1.55 and −1.56: $\frac{-1.55+(-1.56)}{2}=-1.555$.

17. Refer to figure (a).
Find *z* so that the area *A* to the left of *z* is 55% = 0.55. Since *A* = 0.55 > 0.5000, look for a positive *z* value. The
area to the left of 0.13 is 0.5517, so *z* = 0.13.

19. Refer to figure (b).
Find *z* so that the area *A* to the right of *z* is 8% = 0.08. Since *A* to the right of *z* is 0.08,
1 − *A* = 1 − 0.08 = 0.92 is to the left of *z* value. The area to the left of 1.41 is 0.9207.

21. Refer to figure (b).
Find *z* so that the area *A* to the right of *z* is 82% = 0.82. Since *A* to the right of *z*, 1 − *A* = 1 − 0.82 = 0.18 is to
the left of *z*. Since 1 − *A* = 0.18 < 0.5000, look for a negative *z* value. The area to the left of *z* = −0.92 is 0.1788.

23. Refer to figure (c).
Find *z* such that the area between −*z* and *z* is 98% = 0.98. Since *A* is between −*z* and *z*, 1 − *A* = 1 − 0.98 = 0.02
lies in the tails, and since we need ± *z*, half of 1 − *A* lies in each tail.
The area to the left of −*z* is $\frac{1-A}{2}=\frac{0.02}{2}=0.01$.
The area to the left of −2.33 is 0.0099. Thus −*z* = −2.33 and *z* = 2.33.

25. x is approximately normal with $\mu = 85$ and $\sigma = 25$. Since $z = \dfrac{x - \mu}{\sigma}$, we have $z = \dfrac{x - 85}{25}$.

  (a) $P(x > 60) = P\left(\dfrac{x - 85}{25} > \dfrac{60 - 85}{25}\right) = P(z > -1) = 1 - P(z \le -1) = 1 - 0.1587 = 0.8413$

  (b) $P(x < 110) = P\left(\dfrac{x - 85}{25} < \dfrac{110 - 85}{25}\right) = P(z < 1) = 0.8413$

  (c) $P(60 < x < 110) = P(-1 < z < 1) = P(z < 1) - P(z \le -1) = 0.8413 - 0.1587 = 0.6826$

  (i.e., approximately 68% of the blood glucose measurements lies within $\mu \pm \sigma$.)

  (d) $P(x > 140) = P\left(\dfrac{x - 85}{25} > \dfrac{140 - 85}{25}\right) = P(z > 2.2) = 1 - P(z \le 2.2) = 1 - 0.9861 = 0.0139$

27. Pot shard thickness x is approximately normally distributed with $\mu = 5.1$ and $\sigma = 0.9$ millimeters.

  (a) $P(x < 3.0) = P\left(\dfrac{x - 5.1}{0.9} < \dfrac{3.0 - 5.1}{0.9}\right) = P(z < -2.33) = 0.0099$

  (b) $P(x > 7.0) = P\left(\dfrac{x - 5.1}{0.9} > \dfrac{7.0 - 5.1}{0.9}\right) = P(z > 2.11) = 1 - P(z \le 2.11) = 1 - 0.9826 = 0.0174$

  (c) $P(3.0 \le x \le 7.0) = P(-2.33 \le z \le 2.11) = P(z \le 2.11) - P(z < -2.33) = 0.9826 - 0.0099 = 0.9727$

29. Lifetime x is normally distributed with $\mu = 45$ and $\sigma = 8$ months.

  (a) $P(x \le 36) = P\left(\dfrac{x - 45}{8} \le \dfrac{36 - 45}{8}\right) = P(z \le -1.125) \approx P(z \le -1.13) = 0.1292$

  The company will have to replace approximately 13% of its batteries.

  (b) Find x_0 such that $P(x \le x_0) = 10\% = 0.10$. First, find z_0 such that $P(z \le z_0) = 0.10$.
  $P(z \le -1.28) = 0.1003$, so $z_0 = -1.28$. Then $x_0 = \mu + z_0\sigma = 45 + (-1.28)(8) = 34.76 \approx 35$.
  The company should guarantee the batteries for 35 months.

31. Age at replacement x is approximately normal with $\mu = 8$ and range $= 6$ years.
  (a) The empirical rule says that about 95% of the data are between $\mu - 2\sigma$ and $\mu + 2\sigma$, or about 95% of the data are in a $(\mu + 2\sigma) - (\mu - 2\sigma) = 4\sigma$ range (centered around μ). Thus the range $\approx 4\sigma$, or $\sigma \approx$ range/4. Here, we can approximate σ by $6/4 = 1.5$ years.

  (b) $P(x > 5) = P\left(z > \dfrac{5 - 8}{1.5}\right) = P(z > -2)$ Using the estimate of σ from (a)

  $\qquad = 1 - P(z \le -2) = 1 - 0.0228 = 0.9772$

  (c) $P(x < 10) = P\left(z < \dfrac{10 - 8}{1.5}\right) = P(z < 1.33) = 0.9082$

  (d) Find x_0 so that $P(x \le x_0) = 10\% = 0.10$. First, find z_0 such that $P(z \le z_0) = 0.10$.
  $P(z \le -1.28) = 0.1003$, so $z_0 = -1.28$. Then $x_0 = \mu + z_0\sigma = 8 + (-1.28)(1.5) = 6.08$.
  The company should guarantee its TVs for about 6.1 years.

33. Resting heart rate x is approximately normal with $\mu = 46$ and (95%) range from 22 to 70 beats per minute.
  (a) From Problem 31(a), range $\approx 4\sigma$, or $\sigma \approx$ range/4.
  Here, range $= 70 - 22 = 48$, so $\sigma \approx 48/4 = 12$ beats per minute.

  (b) $P(x < 25) = P\left(z < \dfrac{25 - 46}{12}\right) = P(z < -1.75) = 0.0401$

(c) $P(x > 60) = P\left(z > \dfrac{60-46}{12}\right) = P(z > 1.17) = 1 - P(z \le 1.17) = 1 - 0.8790 = 0.1210$

(d) $P(25 \le x \le 60) = P(-1.75 \le z \le 1.17) = P(z \le 1.17) - P(z < -1.75) = 0.8790 - 0.0401 = 0.8389$

(e)   Find x_0 such that $P(x > x_0) = 10\% = 0.10$. First, find z_0 such that $P(z > z_0) = 0.10$.

       $P(z \le z_0) = 1 - 0.10 = 0.90$

       $P(z \le 1.28) = 0.8997 \approx 0.90$, so let $z_0 = 1.28$.

       When $x_0 = \mu + z_0$, $\sigma = 46 + 1.28(12) = 61.36$, so horses with resting rates of 61 beats per minute or more may need treatment.

35. Life expectancy x is normal with $\mu = 90$ and $\sigma = 3.7$ months.

   (a)   The insurance company wants 99% of the microchips to last *longer* than x_0. Thus find x_0 such that

       $P(x \le x_0) = 1\% = 0.01$. First, find z_0 such that $P(z \le z_0) = 0.01$. $P(z \le -2.33) = 0.0099 \approx 0.01$, so let

       $z_0 = -2.33$. Since $z_0 = \dfrac{x_0 - \mu}{\sigma}$, $x_0 = \mu + z_0\sigma = 90 + (-2.33)(3.7) = 81.379 \approx 81$ months.

   (b) $P(x \le 84) = P\left(z \le \dfrac{84-90}{3.7}\right) = P(z \le -1.62) = 0.0526 \approx 5\%$.

   (c)   The "expected loss" is 5.26% [from (b)] of the $50 million, or $0.0526(50,000,000) = \$2,630,000$.

   (d)   Profit is the difference between the amount of money taken in (here, $3 million) and the amount paid out [$2.63 million, from (c)]. Thus the company expects to profit $3,000,000 - 2,630,000 = \$370,000$.

37. Arrival times are normal with $\mu = 3$ hours, 48 minutes and $\sigma = 52$ minutes after the doors open. Convert μ to minutes: $(3 \times 60) + 48 = 228$ minutes.

   (a)   Find x_0 such that $P(x \le x_0) = 90\% = 0.90$. First, find z_0 such that $P(z \le z_0) = 0.90$.

       $P(z \le 1.28) = 0.8997 \approx 0.90$, so let $z_0 = 1.28$.

       Since $z_0 = \dfrac{x_0 - \mu}{\sigma}$, $x_0 = \mu + z_0\sigma = 228 + (1.28)(52) = 294.56$ minutes, or

       $294.56/60 = 4.9093 \approx 4.9$ hours after the doors open.

   (b)   Find x_0 such that $P(x \le x_0) = 15\% = 0.15$. First, find z_0 such that $P(z \le z_0) = 0.15$.

       $P(z \le -1.04) = 0.1492 \approx 0.15$, so let $z_0 = -1.04$. Then

       $x_0 = \mu + z_0\sigma = 228 + (-1.04)(52) = 173.92$ minutes, or $173/60 = 2.899 \approx 2.9$ hours after the doors open.

   (c)   Answers vary. Most people have Saturday off, so many may come early in the day. Most people work Friday, so most people would probably come after 5 P.M. There is no reason to think weekday and weekend arrival times would have the same distribution.

39. Waiting time x is approximately normal with $\mu = 18$ and $\sigma = 4$ minutes.

   (a)   Let A be the event that $x > 20$ and B be the event that $x > 15$. We want to find $P(A, \text{given } B)$.

       Recall $P(A \mid B) = \dfrac{P(A \text{ and } B)}{P(B)}$.

       $P(A \text{ and } B) = P(x > 20 \text{ and } x > 15) = P(x > 20)$

       The number 20 is not included in "both A and B" because A says x is strictly greater than 20. The intervals $(15, \infty)$ and $(20, \infty)$ intersect at $(20, \infty)$.

       $P(x > 20) = P\left(z > \dfrac{20-18}{4}\right) = P(z > 0.5) = 1 - P(z \le 0.5) = 1 - 0.6915 = 0.3085$

$$P(x > 15) = P\left(z > \frac{15-18}{4}\right) = P(z > -0.75) = 1 - P(z \le -0.75) = 1 - 0.2266 = 0.7734$$

$$P(x > 20, \text{ given } x > 15) = \frac{P(x > 20 \text{ and } x > 15)}{P(x > 15)} = \frac{P(x > 20)}{P(x > 15)} = \frac{0.3085}{0.7734} = 0.3989$$

(b) $P(x > 25, \text{ given } x > 18) = \dfrac{P(x > 25 \text{ and } x > 18)}{P(x > 18)} = \dfrac{P(x > 25)}{P(x > 18)} = \dfrac{P\left(z > \frac{25-18}{4}\right)}{P\left(z > \frac{18-18}{4}\right)}$

$$= \frac{P(z > 1.75)}{P(z > 0)} = \frac{1 - P(z \le 1.75)}{1 - P(z \le 0)} = \frac{1 - 0.9599}{1 - 0.5000} = \frac{0.0401}{0.5} = 0.0802$$

(c) Let event $A = x > 20$ and event $B = x > 15$ in the formula.

Section 6.4

1. Answers vary. Students should identify the individuals (subjects) and variable involved, for example, the population of all ages of all people in Colorado, the population of weights of all students in your school, or the population count of all antelope in Wyoming.

3. A population parameter is a numerical descriptive measure of a population, such as μ, the population mean, σ, the population standard deviation, σ^2, the population variance, p, the population proportion, the population maximum and minimum, etc.

5. Statistical inference refers to making conclusions about the value of a population parameter based on information from the corresponding sample statistic and the associated probability distributions. We will do both estimation and testing.

7. They help us visualize the sampling distribution by using tables and graphs that approximately represent the sampling distribution.

9. We studied the sampling distribution of mean trout lengths based on samples of size 5. Other such sampling distributions abound. Notice that the sample size remains the same for each sample in a sampling distribution.

Section 6.5

1. The standard error is the standard deviation for a sampling distribution.

3. $\bar{x}$ is an unbiased estimator for μ, and $\hat{p}$ is an unbiased estimator for p.

5. **(a)** Since $n = 64$, and we don't have any information about the distribution of x, then we can say $\bar{x}$ will have an approximately normal distribution.

$$\mu_{\bar{x}} = \mu_x = 8 \text{ and } \sigma_{\bar{x}} = \frac{\sigma_x}{\sqrt{n}} = \frac{16}{\sqrt{64}} = 2$$

(b) $z = \dfrac{9-8}{2} = 0.50$

(c) $P(\bar{x} > 9) = P\left(z > \dfrac{9-8}{2}\right) = P(z > 0.50) = 1 - P(z < 0.50) = 1 - 0.6915 = 0.3085$

(d) No, since the probability of this happening is 30.85%.

7. **(a)** The required sample size is $n \geq 30$.
   (b) No. If the original distribution is normal, then $\bar{x}$ is distributed normally for any sample size.

9. The distribution with $n = 225$ will have a smaller standard error. Since $\sigma_{\bar{x}} = \dfrac{\sigma}{\sqrt{n}}$, dividing by the square root of 225 will result in a small standard error regardless of the value of σ.

11. **(a)** $\mu_{\bar{x}} = \mu = 15$

$$\sigma_{\bar{x}} = \frac{\sigma}{\sqrt{n}} = \frac{14}{\sqrt{49}} = 2.0$$

Because $n = 49 \geq 30$, by the central limit theorem, we can assume that the distribution of $\bar{x}$ is approximately normal.

$$z = \frac{\bar{x} - \mu}{\sigma_{\bar{x}}} = \frac{\bar{x} - 15}{2.0}$$

$\bar{x} = 15$ converts to $z = \dfrac{15 - 15}{2.0} = 0$

$\bar{x} = 17$ converts to $z = \dfrac{17 - 15}{2.0} = 1$

$$P(15 \leq \bar{x} \leq 17) = P(0 \leq z \leq 1) = P(z \leq 1) - P(z \leq 0) = 0.8413 - 0.5000 = 0.3413$$

(b) $\mu_{\bar{x}} = \mu = 15$

$$\sigma_{\bar{x}} = \frac{\sigma}{\sqrt{n}} = \frac{14}{\sqrt{64}} = 1.75$$

Because $n = 64 \geq 30$, by the central limit theorem, we can assume that the distribution of $\bar{x}$ is approximately normal.

$$z = \frac{\bar{x} - \mu}{\sigma_{\bar{x}}} = \frac{\bar{x} - 15}{1.75}$$

$\bar{x} = 15$ converts to $z = \dfrac{15 - 15}{1.75} = 0$

$\bar{x} = 17$ converts to $z = \dfrac{17 - 15}{1.75} = 1.14$

$$P(15 \leq \bar{x} \leq 17) = P(0 \leq z \leq 1.14) = P(z \leq 1.14) - P(z \leq 0) = 0.8729 - 0.5000 = 0.3729$$

(c) The standard deviation of part (b) is smaller because of the larger sample size. Therefore, the distribution about $\mu_{\bar{x}}$ is narrower.

13. **(a)** $\mu = 75$, $\sigma = 0.8$

$$P(x < 74.5) = P\left(z < \frac{74.5 - 75}{0.8}\right) = P(z < -0.63) = 0.2643$$

(b) $\mu_{\bar{x}} = 75$, $\sigma_{\bar{x}} = \dfrac{\sigma}{\sqrt{n}} = \dfrac{0.8}{\sqrt{20}} = 0.179$

$$P(\bar{x} < 74.5) = P\left(z < \frac{74.5 - 75}{0.179}\right) = P(z < -2.79) = 0.0026$$

(c) No. If the weight of only one car were less than 74.5 tons, we could not conclude that the loader is out of adjustment. If the mean weight for a sample of 20 cars were less than 74.5 tons, we would suspect that the loader is malfunctioning because the probability of this event occurring is 0.26% if indeed the distribution is correct.

15. **(a)** $\mu = 85$, $\sigma = 25$

$$P(x < 40) = P\left(z < \frac{40-85}{25}\right) = P(z < -1.8) = 0.0359$$

(b) The probability distribution of $\bar{x}$ is approximately normal with $\mu_{\bar{x}} = 85$; $\sigma_{\bar{x}} = \frac{\sigma}{\sqrt{n}} = \frac{25}{\sqrt{2}} = 17.68$.

$$P(\bar{x} < 40) = P\left(z < \frac{40-85}{17.68}\right) = P(z < -2.55) = 0.0054$$

(c) $\mu_{\bar{x}} = 85$, $\sigma_{\bar{x}} = \frac{\sigma}{\sqrt{n}} = \frac{25}{\sqrt{3}} = 14.43$

$$P(\bar{x} < 40) = P\left(z < \frac{40-85}{14.43}\right) = P(z < -3.12) = 0.0009$$

(d) $\mu_{\bar{x}} = 85$, $\sigma_{\bar{x}} = \frac{\sigma}{\sqrt{n}} = \frac{25}{\sqrt{5}} = 11.2$

$$P(\bar{x} < 40) = P\left(z < \frac{40-85}{11.2}\right) = P(z < -4.02) < 0.0002$$

(e) Yes. The more tests a patient completes, the stronger is the evidence for excess insulin. If the average value based on five tests were less than 40, the patient is almost certain to have excess insulin.

17. **(a)** $\mu = 63.0$, $\sigma = 7.1$

$$P(x < 54) = P\left(z < \frac{54-63.0}{7.1}\right) = P(z < -1.27) = 0.1020$$

(b) The expected number undernourished is $2{,}200 \times 0.1020 = 224.4$, or about 224.

(c) $\mu_{\bar{x}} = 63.0$, $\sigma_{\bar{x}} = \frac{\sigma}{\sqrt{n}} = \frac{7.1}{\sqrt{50}} = 1.004$

$$P(\bar{x} < 60) = P\left(z < \frac{60-63.0}{1.004}\right) = P(z < -2.99) = 0.0014$$

(d) $\mu_{\bar{x}} = 63.0$, $\sigma_{\bar{x}} = 1.004$

$$P(\bar{x} < 64.2) = P\left(z < \frac{64.2-63.0}{1.004}\right) = P(z < 1.20) = 0.8849$$

Since the sample average is above the mean, it is quite unlikely that the doe population is undernourished.

19. **(a)** The random variable x is itself an average based on the number of stocks or bonds in the fund. Since x itself represents a sample mean return based on a large (random) sample of size $n = 250$ of stocks or bonds, x has a distribution that is approximately normal (central limit theorem).

(b) $\mu_{\bar{x}} = 1.6\%$, $\sigma_{\bar{x}} = \frac{\sigma}{\sqrt{n}} = \frac{0.9\%}{\sqrt{6}} = 0.367\%$

$$P(1\% \leq \bar{x} \leq 2\%) = P\left(\frac{1\%-1.6\%}{0.367\%} \leq z \leq \frac{2\%-1.6\%}{0.367\%}\right) = P(-1.63 \leq z \leq 1.09)$$

$$= P(z \leq 1.09) - P(z \leq -1.63) = 0.8621 - 0.0516 = 0.8105$$

(c) *Note:* 2 years = 24 months; x is *monthly* percentage return.

$$\mu_{\bar{x}} = 1.6\%, \quad \sigma_{\bar{x}} = \frac{\sigma}{\sqrt{n}} = \frac{0.9\%}{\sqrt{24}} = 0.1837\%$$

$$P(1\% \leq \overline{x} \leq 2\%) = P\left(\frac{1\% - 1.6\%}{0.1837\%} \leq z \leq \frac{2\% - 1.6\%}{0.1837\%}\right) = P(-3.27 \leq z \leq 2.18)$$

$$= P(z \leq 2.18) - P(z \leq -3.27) = 0.9854 - 0.0005 = 0.9849$$

(d) Yes. The probability increases as the standard deviation decreases. The standard deviation decreases as the sample size increases.

(e) $\mu_{\overline{x}} = 1.6\%$, $\sigma_{\overline{x}} = 0.1837\%$

$$P(\overline{x} < 1\%) = P\left(z < \frac{1\% - 1.6\%}{0.1837\%}\right) = P(z < -3.27) = 0.0005$$

This is very unlikely if $\mu = 1.6\%$. One would suspect that μ has slipped below 1.6%.

21. **(a)** The total checkout time for 30 customers is the sum of the checkout times for each individual customer. Thus $w = x_1 + x_2 + \cdots + x_{30}$, and the probability that the total checkout time for the next 30 customers is less than 90 is $P(w < 90)$.

(b) If we divide both sides of $w < 90$ by 30, we obtain $\dfrac{w}{30} < 3$. However, w is the sum of 30 waiting times, so

$\dfrac{w}{30}$ is $\overline{x}$. Therefore, $P(w < 90) = P(\overline{x} < 3)$.

(c) The probability distribution of $\overline{x}$ is approximately normal with mean $\mu_{\overline{x}} = \mu = 2.7$ and standard deviation

$$\sigma_{\overline{x}} = \frac{\sigma}{\sqrt{n}} = \frac{0.6}{\sqrt{30}} = 0.1095.$$

(d) $P(\overline{x} < 3) = P\left(z < \dfrac{3 - 2.7}{0.1095}\right) = P(z < 2.74) = 0.9969$

The probability that the total checkout time for the next 30 customers is less than 90 minutes is 0.9969.

23. **(a)** Let $w = x_1 + x_2 + \cdots + x_5$.

$$\mu_{\overline{x}} = \mu = 17, \ \sigma_{\overline{x}} = \frac{\sigma}{\sqrt{n}} = \frac{3.3}{\sqrt{5}} = 1.476$$

$$P(w > 90) = P\left(\frac{w}{5} > \frac{90}{5}\right) = P(\overline{x} > 18) = P\left(z > \frac{18 - 17}{1.476}\right)$$

$$= P(z > 0.68) = 1 - 0.7517 = 0.2483$$

(b) $P(w < 80) = P\left(\dfrac{w}{5} < \dfrac{80}{5}\right) = P(\overline{x} < 16) = P\left(z < \dfrac{16 - 17}{1.476}\right)$

$$= P(z < -0.68) = 0.2483$$

(c) $P(80 < w < 90) = P(16 < \overline{x} < 18) = P(-0.68 < z < 0.68)$

$$= P(z < 0.68) - P(z < -0.68) = 0.7517 - 0.2483 = 0.5034$$

Section 6.6

Answers may vary slightly due to rounding.

1. We need $np > 5$ and $nq > 5$. The approximation improves as n increases.

3. **(a)** Yes. Here, $np = 40(0.5) = 20 > 5$ and $nq = 40(0.5) = 20 > 5$.

    (b) $\mu = np = 40(0.5) = 20$; $\sigma = \sqrt{npq} = \sqrt{40(0.5)(0.5)} \approx 3.162$

    (c) $P(r \geq 23)$ corresponds to $P(x \geq 22.5)$

    (d) $P(r \geq 23) \approx P(x \geq 22.5) = P\left(\dfrac{x - \mu}{\sigma} \geq \dfrac{22.5 - 20}{3.162} \right) =$

    $P(z \geq 0.7906) = 1 - P(z < 0.7906) = 0.2146$

    (e) It is not unusual, as the probability is 21.46%.

5. No. Here, $np = 10 \times 0.43 = 4.3 < 5$. The conditions are not met. Use a TI-83 or software.

7. Previously, $p = 88\% = 0.88$; now, $p = 9\% = 0.09$; $n = 200$; $r = 50$.
    Let a success be defined as a child with a high blood lead level.
    (a) $P(r \geq 50) = P(50 \leq r) = P(49.5 \leq x)$

    $np = 200(0.88) = 176$; $nq = n(1 - p) = 200(0.12) = 24$

    Since both np and nq are greater than 5, we will use the normal approximation to the binomial with

    $\mu = np = 176$ and $\sigma = \sqrt{npq} = \sqrt{200(0.88)(0.12)} = \sqrt{21.12} = 4.60$.

    So, $P(r \geq 50) = P(49.5 \leq x) = P\left(\dfrac{49.5 - 176}{4.6} \leq z \right) = P(-27.5 \leq z)$.

    Almost every z value will be greater than or equal to -27.5, so this probability is approximately 1. It is almost certain that 50 or more children had high blood lead levels a decade ago.

    (b) $P(r \geq 50) = P(50 \leq r) = P(49.5 \leq x)$

    In this case, $np = 200(0.09) = 18$ and $nq = 200(0.91) = 182$, so both are greater than 5. Use the normal

    approximation with $\mu = np = 18$ and $\sigma = \sqrt{npq} = \sqrt{200(0.09)(0.91)} = \sqrt{16.38} = 4.05$.

    So $P(49.5 \leq x) = P\left(\dfrac{49.5 - 18}{4.05} \leq z \right) = P(7.78 \leq z)$.

    Almost no z values will be larger than 7.78, so this probability is approximately 0. Today, it is almost impossible that a sample of 200 children would include at least 50 with high blood lead levels.

9. We are given $n = 753$ and $p = 3.5\% = 0.035$; $q = 1 - p = 1 - 0.035 = 0.965$.
    Let a success be a person living past age 90.

    (a) $P(r \geq 15) = P(15 \leq r) = P(14.5 \leq x)$ 15 is a left endpoint.

    Here, $np = 753(0.035) = 26.355$, and $nq = 753(0.965) = 726.645$, both of which are greater than 5; the normal approximation is appropriate, using $\mu = np = 26.355$ and

    $\sigma = \sqrt{npq} = \sqrt{753(0.035)(0.965)} = \sqrt{25.4326} = 5.0431$.

    $P(14.5 \leq x) = P\left(\dfrac{14.5 - 26.355}{5.0431} \leq z \right) = P(-2.35 \leq z)$

    $= P(z \geq -2.35) = 1 - P(z < -2.35) = 1 - 0.0094 = 0.9906$

(b) $P(r \geq 30) = P(30 \leq r) = P(29.5 \leq x) = P\left(\dfrac{29.5 - 26.355}{5.0431} \leq z\right)$

$\qquad = P(0.62 \leq z) = P(z \geq 0.62) = 1 - P(z < 0.62) = 1 - 0.7324 = 0.2676$

(c) $P(25 \leq r \leq 35) = P(24.5 \leq x \leq 35.5) = P\left(\dfrac{24.5 - 26.355}{5.0431} \leq z \leq \dfrac{35.5 - 26.355}{5.0431}\right)$

$\qquad = P(-0.37 \leq z \leq 1.81) = P(z \leq 1.81) - P(z < -0.37) = 0.9649 - 0.3557 = 0.6092$

(d) $P(r > 40) = P(r \geq 41) = P(41 \leq r) = P(40.5 \leq x) = P\left(\dfrac{40.5 - 26.355}{5.0431} \leq z\right)$

$\qquad = P(2.80 \leq z) = P(z \geq 2.80) = 1 - P(z < 2.80) = 1 - 0.9974 = 0.0026$

11. $n = 66, p = 80\% = 0.80, q = 1 - p = 1 - 0.80 = 0.20$
A success is when a new product fails within 2 years.

(a) $P(r \geq 47) = P(47 \leq r) = P(46.5 \leq x)$

$np = 66(0.80) = 52.8$, and $nq = 66(0.20) = 13.3$. Both exceed 5, so the normal approximation with
$\mu = np = 52.8$ and $\sigma = \sqrt{npq} = \sqrt{66(0.8)(0.2)} = \sqrt{10.56} = 3.2496$ is appropriate.

$P(46.5 \leq x) = P\left(\dfrac{46.5 - 52.8}{3.2496} \leq z\right) = P(-1.94 \leq z)$

$\qquad = P(z \geq -1.94) = 1 - P(z < -1.94) = 1 - 0.0262 = 0.9738$

(b) $P(r \leq 58) = P(x \leq 58.5) = P\left(z \leq \dfrac{58.5 - 52.8}{3.2496}\right) = P(z \leq 1.75) = 0.9599$

For (c) and (d), note that we are now interested in products succeeding, so a success is redefined to be a
new product staying on the market for 2 years. Here, $n = 66$, p is now 0.20, and q now is 0.80 (p and q
above have been switched). Now $np = 13.2$ and $nq = 52.8$, $\mu = 13.2$, and σ stays equal to 3.2496.

(c) $P(r \geq 15) = P(15 \leq r) = P(14.5 \leq x) = P\left(\dfrac{14.5 - 13.2}{3.2496} \leq z\right)$

$\qquad = P(0.40 \leq z) = P(z \geq 0.40) = 1 - P(z < 0.40) = 1 - 0.6554 = 0.3446$

(d) $P(r < 10) = P(r \leq 9) = P(x \leq 9.5) = P\left(z \leq \dfrac{9.5 - 13.2}{3.2496}\right) = P(z \leq -1.14) = 0.1271$

13. $n = 317$, P(buy, given sampled) $= 37\% = 0.37$, P(sampled) $= 60\% = 0.60 = p$ so $q = 0.40$.

(a) $P(180 < r) = P(181 \leq r) = P(180.5 \leq x)$

$np = 190.2$, $nq = 126.8$, $\sigma = \sqrt{npq} = \sqrt{76.08} = 8.7224$

Since both np and np are greater than 5, use normal approximation with $\mu = np$ and $\sigma = \sqrt{npq}$.

$P(180.5 \leq x) = P\left(\dfrac{180.5 - 190.2}{8.7224} \leq z\right) = P(-1.11 \leq z) = 1 - P(z < -1.11) = 1 - 0.1335 = 0.8665$

(b) $P(r < 200) = P(r \leq 199) = P(x \leq 199.5) = P\left(z \leq \dfrac{199.5 - 190.2}{8.7224}\right) = P(z \leq 1.07) = 0.8577$

(c) Let A be the event buy product, and let B be the event tried free sample. Thus $P(A \mid B) = 0.37$, and $P(B) = 0.60$. Since $P(A \text{ and } B) = P(B) \times P(A \mid B) = 0.60 \times 0.37 = 0.222$.
Thus $P(\text{sample and buy}) = 0.222$.

(d) Let a success be sample and buy. Then $p = 0.222$ from (c), and $q = 0.778$.
$P(60 \le r \le 80) = P(59.5 \le x \le 80.5)$

Here, $np = 317(0.222) = 70.374$ and $nq = 246.626$, so use normal approximation with $\mu = np$ and
$\sigma = \sqrt{npq} = \sqrt{317(0.222)(0.778)} = \sqrt{54.750972} = 7.3994$.

$$P(59.5 \le x \le 80.5) = P\left(\frac{59.5 - 70.374}{7.3994} \le z \le \frac{80.5 - 70.374}{7.3994} \right) = P(-1.47 \le z \le 1.37)$$
$$= P(z \le 1.37) - P(z < -1.47) = 0.9147 - 0.0708 = 0.8439$$

15. $n = 267$ reservations, $P(\text{show}) = 1 - 0.06 = 0.94 = p$ so $q = 0.06$.

   (a) $p = 0.94$

   (b) Success = show up for flight (with a reservation) seat available for all who show up means the number showing up must be ≤ 255 actual plane seats. Thus $P(r \le 255)$.

   (c) $P(r \le 255) = P(x \le 255.5)$
   Since $np = 267(0.94) = 250.98 > 5$ and $nq = 267(0.06) = 16.02 > 5$, use a normal approximation with
   $\mu = np$ and $\sigma = \sqrt{npq} = \sqrt{267(0.94)(0.06)} = \sqrt{15.0588} = 3.8806$.

   $$P(x \le 255.5) = P\left(z \le \frac{255.5 - 250.98}{3.8806} \right) = P(z \le 1.16) = 0.8770$$

17. When $np > 5$ and $nq > 5$.

19. Yes, $\hat{p}$ is an unbiased estimator for p under the conditions $np > 5$ and $nq > 5$.

21. (a) Yes we can use the approximation since both $np = 100(0.23) = 23 > 5$ and
   $nq = 100(0.77) = 77 > 5$. Here, $\mu_{\hat{p}} = p = 0.23$ and $\sigma_{\hat{p}} = \sqrt{\dfrac{pq}{n}} = \sqrt{\dfrac{0.23(0.77)}{100}} \approx 0.042$.

   (b) No, since $np = 20(0.23) = 4.6 < 5$.

Chapter Review Problems

1. Normal probability distributions model continuous random variables. They are symmetric around their mean, are bell-shaped, and extend across the entire real number line. However, most of the data fall within 3 standard deviations from the mean. The mean, median, and mode are all equal.

3. No, this violates one of the three conditions.

5. The points on the plot must lie close to a straight line.

7. $\sigma_{\bar{x}} = \dfrac{\sigma_x}{\sqrt{n}}$

9. **(a)** The $\bar{x}$ distribution approaches a normal distribution.

   (b) The mean $\mu_{\bar{x}}$ of the $\bar{x}$ distribution equals the mean μ of the x distribution regardless of the sample size.

   (c) The standard deviation $\sigma_{\bar{x}}$ of the sampling distribution equals $\dfrac{\sigma}{\sqrt{n}}$, where σ is the standard deviation of the x distribution and n is the sample size.

   (d) They will both be approximately normal with the same mean, but the standard deviations will be $\dfrac{\sigma}{\sqrt{50}}$ and $\dfrac{\sigma}{\sqrt{100}}$, respectively.

11. x is normal with $\mu = 47$ and $\sigma = 6.2$.

   (a) $P(x \le 60) = P\left(z \le \dfrac{60-47}{6.2}\right) = P(z \le 2.10) = 0.9821$

   (b) $P(x \ge 50) = P\left(z \ge \dfrac{50-47}{6.2}\right) = P(z \ge 0.48) = 1 - P(z < 0.48) = 1 - 0.6844 = 0.3156$

   (c) $P(50 \le x \le 60) = P(0.48 \le z \le 2.10) = P(z \le 2.10) - P(z < 0.48) = 0.9821 - 0.6844 = 0.2977$

13. Find z_0 such that $P(z \ge z_0) = 5\% = 0.05$. Same as find z_0 such that $P(z < z_0) = 0.95$.
    $P(z < 1.645) = 0.95$, so $z_0 = 1.645$.

15. Success = having blood type AB
    $n = 250, p = 3\% = 0.03, q = 1 - p = 0.97, np = 7.5, nq = 242.5, \sqrt{npq} = 2.6972$

   (a) $P(5 \le r) = P(4.5 \le x)$

   $np > 7.5$ and $\sigma = \sqrt{npq} = \sqrt{7.275} = 2.6972$

   $P(4.5 \le x) = P\left(\dfrac{4.5-7.5}{2.6972} \le z\right) = P(-1.11 \le z) = 1 - P(z < -1.11) = 1 - 0.1335 = 0.8665$

   (b) $P(5 \le r \le 10) = P(4.5 \le x \le 10.5) = P\left(-1.11 \le z \le \dfrac{10.5-7.5}{2.6972}\right)$

   $= P(-1.11 \le z \le 1.11) = 1 - 2P(z < -1.11) = 1 - 2(0.1335) = 0.7330$

17. Binomial with $n = 400, p = 0.70,$ and $q = 0.30$.
    Success = can recycled

   (a) $P(r \ge 300) = P(300 \le r) = P(299.5 \le x)$

   $np = 280 > 5, nq = 120 > 5, \sqrt{npq} = \sqrt{84} = 9.1652$

   Use normal approximation with $\mu = np$ and $\sigma = \sqrt{npq}$.

   $P(299.5 \le x) = P\left(\dfrac{299.5-280}{9.1652} \le z\right) = P(2.13 \le z) = 1 - P(z < 2.13) = 1 - 0.9834 = 0.0166$

(b) $P(260 \le r \le 300) = P(259.5 \le x \le 300.5) = P\left(\dfrac{259.5 - 280}{9.1652} \le z \le \dfrac{300.5 - 280}{9.1652}\right)$

$$= P(-2.24 \le z \le 2.24) = P(z \le 2.24) - P(z < -2.24) = 0.9875 - 0.0125 = 0.9750$$

19. Delivery time x is normal with $\mu = 14$ and $\sigma = 2$ hours.

   (a) $P(x \le 18) = P\left(z \le \dfrac{18 - 14}{2}\right) = P(z \le 2) = 0.9772$

   (b) Find x_0 such that $P(x \le x_0) = 0.95$.

   Find z_0 such that $P(z \le z_0) = 0.95$.

   $P(z \le 1.645) = 0.95$, so $z_0 = 1.645$.

   $x_0 = \mu + z_0 \sigma = 14 + 1.645(2) = 17.29 \approx 17.3$ hours

21. **(a)** $\mu = 35,\ \sigma = 7$

   $$P(x \ge 40) = P\left(z \ge \dfrac{40 - 35}{7}\right) = P(z \ge 0.71) = 0.2389$$

   (b) $\mu_{\bar{x}} = \mu = 35,\ \sigma_{\bar{x}} = \dfrac{\sigma}{\sqrt{n}} = \dfrac{7}{\sqrt{9}} = \dfrac{7}{3}$

   $$P(\bar{x} \ge 40) = P\left(z \ge \dfrac{40 - 35}{\frac{7}{3}}\right) = P(z \ge 2.14) = 0.0162$$

23. $\mu_{\bar{x}} = \mu = 100,\ \sigma_{\bar{x}} = \dfrac{\sigma}{\sqrt{n}} = \dfrac{15}{\sqrt{100}} = 1.5$

   $P(100 - 2 \le \bar{x} \le 100 + 2) = P(98 \le \bar{x} \le 102) = P\left(\dfrac{98 - 100}{1.5} \le z \le \dfrac{102 - 100}{1.5}\right)$

   $$= P(-1.33 \le z \le 1.33) = P(z \le 1.33) - P(z \le -1.33) = 0.9082 - 0.0918 = 0.8164$$

25. **(a)** Yes, as both $np = 24(0.4) = 9.6 > 5$ and $nq = 24(0.6) = 14.4 > 5$.

   (b) $\mu_{\hat{p}} = p = 0.4$; $\sigma_{\hat{p}} = \sqrt{\dfrac{pq}{n}} = \sqrt{\dfrac{0.4(0.6)}{24}} = 0.1$

Cumulative Review Problems Chapters 4, 5, 6

1. A statistical experiment entails the occurrence of trials with outcomes, either measured or counted. Each outcome has an associated probability, and usually there is a probability distribution, such as the normal, binomial, or Poisson, for each statistical experiment. For this example, magnetic susceptibility can take on positive values, so the intervals listed cover all possible positive values. Therefore, it is a listing of the sample space.

3. Yes, they do. Since we have a listing of the sample space, the probabilities must add to 1.

5. $P(40 \leq x \mid 20 \leq x) = \dfrac{P(40 \leq x \text{ and } 20 \leq x)}{P(20 \leq x)} = \dfrac{P(40 \leq x)}{P(20 \leq x)} = \dfrac{0.05}{0.15 + 0.10 + 0.05} = 0.167$

7. **(a)** Let a success be "very interesting." Then $p = 0.10$, $q = 0.90$, and $n = 12$.

   (b) $\mu = np = 12 \times 0.10 = 1.2$

   $\sigma = \sqrt{npq} = \sqrt{12 \times 0.10 \times 0.90} \approx 1.04$

   (c) $P(r \geq 1) = 1 - P(r = 0) = 1 - \left[1 \times (0.10)^0 \times (0.90)^{12} \right] = 1 - 0.282 = 0.718$

   (d) $P(r < 3) = P(r = 0) + P(r = 1) + P(r = 2) = 0.2825 + 0.3766 + 0.2301 = 0.8892$

9. **(a)** Yes; since $n = 100$ and $np = 5$, the criteria are satisfied. Let $\lambda = 5$.
   (b) By Table 4 in Appendix II, $P(x \leq 6) = 0.7622$.
   (c) By Table 4 in Appendix II, $P(x \geq 8) = 1 - P(x \leq 7) = 1 - 0.932 = 0.068$.

11. **(a)** $\sigma \approx \dfrac{13.3 - 6.5}{4} = 1.7$

    (b)
    $P(x < 8) = P(z < -1.12) = 0.1314$

    $z = \dfrac{8 - 9.9}{1.7} = -1.12$

    (c)
    $P(x > 12) = P(z > 1.24) = 0.1075$

    $z = \dfrac{12 - 9.9}{1.7} = 1.24$

13. **(a)** Since the sample size is large, the central limit theorem describes the distribution of $\bar{x}$.
    (b)
    $P(\bar{x} \leq 6820) = P(z \leq -2.75) = 0.0030$

    $z = \dfrac{6820 - 7500}{1750 / \sqrt{50}} = -2.75$

    (c) The probability that the 50 workers' average white blood cell count is that low or lower, if they were healthy adults, is extremely low. It would be reasonable to gather additional facts.

15. Answers vary. The normal distribution describes continuous measurements such as heights, weights, and profits; e.g., this distribution is bell-shaped and symmetric. The binomial distribution describes counts of experiments that have exactly two outcomes. This distribution describes things such as number of females in a class, number of cavities in a mouth, or number of heads shown when flipping coins. The Poisson distribution describes the number of occurrences of rare events, such as number of diseased trees in a forest or the number of bombs hitting their targets. The binomial and Poisson distributions are for discrete random variables. We can approximate the binomial distribution with the normal when np and nq exceed 5. We can approximate the binomial distribution with the Poisson if n exceeds 100 but np is less than 10.

Chapter 7: Estimation

Section 7.1

Answers may vary slightly owing to rounding.

1. True. By definition, critical values z_c are such that $c\%$ of the area under the standard normal curve falls between $-z_c$ and z_c.

3. True. By definition, the margin of error is the magnitude of the difference between $\bar{x}$ and μ.

5. False. The maximum error rate is $E = z_c \dfrac{\sigma}{\sqrt{n}}$. As the sample size n increases, the maximal error decreases, resulting in a shorter confidence interval for μ.

7. False. The maximal error of estimate E controls the length of the confidence interval regardless of the value of $\bar{x}$.

9. Either μ is contained in the interval or it is not. Therefore, the probability that μ is in this interval is 0 or 1, not 0.95.

11. **(a)** Yes. The x distribution is normal, so the $\bar{x}$ distribution is also normal and σ is known.

    (b)
    $$\bar{x} - E < \mu < \bar{x} + E$$
    $$\bar{x} - z_c \frac{\sigma}{\sqrt{n}} < \mu < \bar{x} + z_c \frac{\sigma}{\sqrt{n}}$$
    $$50 - 1.645\left(\frac{6}{\sqrt{16}}\right) < \mu < 50 + 1.645\left(\frac{6}{\sqrt{16}}\right)$$
    $$50 - 2.468 < \mu < 50 + 2.468$$
    $$47.532 < \mu < 52.468$$

    (c) We are 90% confident that this interval contains μ from this population.

13. **(a)** $n = \left(\dfrac{z_c \sigma}{E}\right)^2 = \left(\dfrac{1.96 \times 3}{0.4}\right)^2 = 216.09 \rightarrow n = 217$

    (b) Yes, by the central limit theorem $\bar{x}$ will have an approximate normal distribution.

15. **(a)** $n = 15$, $\bar{x} = 3.15$, $\sigma = 0.33$, $c = 80\%$, $z_c = 1.28$
    $$E = \frac{z_c \sigma}{\sqrt{n}} = \frac{1.28(0.33)}{\sqrt{15}} \approx 0.11$$
    $$(\bar{x} - E) < \mu < (\bar{x} + E)$$
    $$3.15 - 0.11 < \mu < 3.15 + 0.11$$
    $$3.04 \text{ g} < \mu < 3.26 \text{ g}$$

    The margin of error E is 0.11 g.

    (b) The distribution of weights is assumed to be normal, and σ is known.

(c) There is an 80% chance this confidence interval is one of the intervals that contains the population average weight of Allen's hummingbirds in this region.

(d) $n = \left(\dfrac{z_c \sigma}{E}\right)^2 = \left(\dfrac{1.28 \times 0.33}{0.08}\right)^2 = 27.88$, so $n = 28$ hummingbirds.

17. (a) $n = 45$, $\bar{x} = 37.5$, $\sigma = 7.50$, $c = 99\%$, $z_c = 2.58$

$$E = \frac{z_c \sigma}{\sqrt{n}} = \frac{2.58(7.50)}{\sqrt{45}} \approx 2.88$$
$$(\bar{x} - E) < \mu < (\bar{x} + E)$$
$$37.5 - 2.88 < \mu < 37.5 + 2.88$$
$$34.62 \text{ mL/kg} < \mu < 40.38 \text{ mL/kg}$$

The margin of error E is 2.88 mL/kg.

(b) The sample size is large (30 or more), and σ is known.

(c) There is a 99% chance this confidence interval is one of the intervals that contains the population average blood plasma level for male firefighters.

(d) $n = \left(\dfrac{z_c \sigma}{E}\right)^2 = \left(\dfrac{2.58 \times 7.50}{2.50}\right)^2 = 59.91$, so $n = 60$ plasma samples.

19. $n = 30$, $\bar{x} = 138.5$, $\sigma = 42.6$

(a) $c = 90\%$, $z_c = 1.645$

$$E = \frac{z_c \sigma}{\sqrt{n}} = \frac{1.645(42.6)}{\sqrt{30}} \approx 12.8$$
$$(\bar{x} - E) < \mu < (\bar{x} + E)$$
$$(138.5 - 12.8) < \mu < (138.5 + 12.8)$$
$$125.7 \text{ larceny cases} < \mu < 151.3 \text{ larceny cases}$$

The margin of error E is 12.8 larceny cases.

(b) $c = 95\%$, $z_c = 1.96$

$$E = \frac{z_c \sigma}{\sqrt{n}} = \frac{1.96(42.6)}{\sqrt{30}} \approx 15.2$$
$$(\bar{x} - E) < \mu < (\bar{x} + E)$$
$$(138.5 - 15.2) < \mu < (138.5 + 15.2)$$
$$123.3 \text{ larceny cases} < \mu < 153.7 \text{ larceny cases}$$

The margin of error E is 15.2 larceny cases.

(c) $c = 99\%$, $z_c = 2.58$

$$E = \frac{z_c \sigma}{\sqrt{n}} = \frac{2.58(42.6)}{\sqrt{30}} \approx 20.1$$
$$(\bar{x} - E) < \mu < (\bar{x} + E)$$
$$(138.5 - 20.1) < \mu < (138.5 + 20.1)$$
$$118.4 \text{ larceny cases} < \mu < 158.6 \text{ larceny cases}$$

The margin of error E is 20.1 larceny cases.

(d) Yes.
(e) Yes.

21. (a)

$$\bar{x} - z_c \frac{\sigma}{\sqrt{n}} < \mu < \bar{x} + z_c \frac{\sigma}{\sqrt{n}}$$

$$30 - 1.96\left(\frac{12}{\sqrt{49}}\right) < \mu < 30 + 1.96\left(\frac{12}{\sqrt{49}}\right)$$

$$30 - 3.36 < \mu < 30 + 3.36$$

$$26.64 < \mu < 33.36$$

The margin of error is 3.36.

(b)

$$\bar{x} - z_c \frac{\sigma}{\sqrt{n}} < \mu < \bar{x} + z_c \frac{\sigma}{\sqrt{n}}$$

$$30 - 1.96\left(\frac{12}{\sqrt{100}}\right) < \mu < 30 + 1.96\left(\frac{12}{\sqrt{100}}\right)$$

$$30 - 2.35 < \mu < 30 + 2.35$$

$$27.65 < \mu < 32.35$$

The margin of error is 2.35.

(c)

$$\bar{x} - z_c \frac{\sigma}{\sqrt{n}} < \mu < \bar{x} + z_c \frac{\sigma}{\sqrt{n}}$$

$$30 - 1.96\left(\frac{12}{\sqrt{225}}\right) < \mu < 30 + 1.96\left(\frac{12}{\sqrt{225}}\right)$$

$$30 - 1.57 < \mu < 30 + 1.57$$

$$28.43 < \mu < 31.57$$

The margin of error is 1.57.

(d) Yes. As the sample size increases, the margin of error decreases.

(e) Yes. As the sample size increases, the length of the 95% confidence intervals decreases.

23. (a) $n = 42$, $\bar{x} = \dfrac{\sum x_i}{n} = \dfrac{1,511.8}{42} = 35.9952 \approx 36.0$, as stated.

(b) $c = 75\%$, $z_c = 1.15$

$$E \approx \frac{z_c \sigma}{\sqrt{n}} = \frac{1.15(10.2)}{\sqrt{42}} = 1.81$$

$$(\bar{x} - E) < \mu < (\bar{x} + E)$$

$$(36.0 - 1.81) < \mu < (36.0 + 1.81)$$

$34.19 < \mu < 37.81$ thousand dollars per employee profit

(c) Yes. Since $30,000 per employee profit is less than the lower limit of the confidence interval, $34,190, your bank profits are low compared with those of other similar financial institutions.

(d) Yes. Since $40,000 per employee profit exceeds the upper limit of the confidence interval, $37,810, your bank profits are higher than those of other similar financial institutions.

(e) $c = 90\%$, $z_c = 1.645$

$$E \approx \frac{z_c \sigma}{\sqrt{n}} = \frac{1.645(10.2)}{\sqrt{42}} = 2.59$$
$$(\bar{x} - E) < \mu < (\bar{x} + E)$$
$$(36.0 - 2.59) < \mu < (36.0 + 2.59)$$

$33.41 < \mu < 38.59$ thousand dollars per employee profit

Yes. $30,000 is less than the lower limit of the confidence interval, $33,410, so your bank profits are less than those of other financial institutions.

Yes. $40,000 is more than the upper limit of the confidence interval, $38,590, so your bank profits are higher than those of other financial institutions.

25.   (a) $c = 95\%$, $z_c = 1.96$

$$E \approx \frac{z_c \sigma}{\sqrt{n}} = \frac{1.96(17)}{\sqrt{56}} = 4.5$$
$$(\bar{x} - E) < \mu < (\bar{x} + E)$$
$$(97 - 4.5) < \mu < (97 + 4.5)$$
$$92.5 < \mu < 101.5 \text{ degrees Celsius}$$

(b)   The balloon should rise with the higher average temperature.

Section 7.2

Answers may vary slightly owing to rounding.

1. $n = 18$, so $d.f. = n - 1 = 18 - 1 = 17$, $c = 0.95$; $t_c = t_{0.95} = 2.110$

3. $n = 22$, so $d.f. = n - 1 = 22 - 1 = 21$, $c = 0.90$; $t_c = t_{0.90} = 1.721$

5.   All t distributions are symmetric around $t = 0$.

7.   As n increases, the value of t_c decreases. Therefore, t_c is larger for $n = 10$ with $d.f. = 9$.

9.   Shorter. For $d.f. = 40$, z_c is less than t_c, and the resulting margin of error E is smaller.

11.   (a)   Yes. The x distribution is mound-shaped and we have s from the sample.

(b) $df = n - 1 = 15$; $t_c = 1.753$

$$\bar{x} - E < \mu < \bar{x} + E$$
$$\bar{x} - t_c \frac{s}{\sqrt{n}} < \mu < \bar{x} + t_c \frac{s}{\sqrt{n}}$$
$$10 - 1.753 \frac{2}{\sqrt{16}} < \mu < 10 + 1.753 \frac{2}{\sqrt{16}}$$
$$10 - 0.877 < \mu < 10 + 0.877$$
$$9.123 < \mu < 10.877$$

(c)   There is a 90% chance that this confidence interval is one of the confidence intervals that contains μ.

13. $n = 9$, so $d.f. = n - 1 = 9 - 1 = 8$.

   (a) $\bar{x} = \dfrac{\sum x}{n} = \dfrac{11,450}{9} \approx 1,272$, as stated

$$s^2 = \frac{\sum x_i^2 - \dfrac{(\sum x_i)^2}{n}}{n-1} = \frac{14,577,854 - \dfrac{(11,450)^2}{9}}{8} = 1,363.6944$$

$$s = \sqrt{1,363.6944} = 36.9282 \approx 37, \text{ as stated}$$

   (b) $c = 90\%$, $t_c = t_{0.90}$ with 8 $d.f. = 1.860$

$$E = \frac{t_c s}{\sqrt{n}} = \frac{1.86(37)}{\sqrt{9}} = 22.94 \approx 23$$

$$(\bar{x} - E) < \mu < (\bar{x} + E)$$

$$(1272 - 23) < \mu < (1272 + 23)$$

$$1249 \text{ A.D.} < \mu < 1295 \text{ A.D.}$$

   (c)   We are 90% confident that the computed interval is one that contains the population mean for the tree-ring date.

15. $n = 6$ so $d.f. = n - 1 = 5$

   (a) $\bar{x} = 91.0$, as stated

           $s = 30.7181 \approx 30.7$, as stated

   (b) $c = 75\%$, so $t_{0.75}$ with 5 $d.f. = 1.301$.

$$E = \frac{t_c s}{\sqrt{n}} = \frac{1.301(30.7)}{\sqrt{6}} \approx 16.3$$

$$(\bar{x} - E) < \mu < (\bar{x} + E)$$

$$(91.0 - 16.3) < \mu < (91.0 + 16.3)$$

$$74.7 \text{ pounds} < \mu < 107.3 \text{ pounds}$$

   (c)   We are 75% confident that the computed interval is one that contains the population mean weight of adult mountain lions in the region.

17. $n = 10$, so $d.f. = n - 1 = 9$.

   (a) $\bar{x} = 9.9500 \approx 9.95$, as stated

           $s \approx 1.0212 \approx 1.02$, as stated

   (b) $c = 99.9\%$, so $t_{0.999}$ with 9 $d.f. = 4.781$.

$$E = \frac{t_c s}{\sqrt{n}} = \frac{4.781(1.02)}{\sqrt{10}} \approx 1.54$$

$$(\bar{x} - E) < \mu < (x + E)$$

$$9.95 - 1.54 < \mu < 9.95 + 1.54$$

$$8.41 \text{ mg/dL} < \mu < 11.49 \text{ mg/dL}$$

   (c)   Since all values in the 99.9% confidence interval are above 6 mg/dL, we can be almost certain that this patient does not have a calcium deficiency.

19. Notice that the four figures are drawn on different scales.

    (a) The boxplots differ in range, interquartile range, median, symmetry, whisker lengths, and the presence or absence of outliers. These differences are to be expected because each boxplot represents a different sample of size 20. Although the data sets all were selected as samples of size $n = 20$ from a normal distribution with $\mu = 68$ and $\sigma = 3$, it is interesting that Sample 2, Figure (b), shows two outliers.

    (b)

Sample	Confidence Interval Width	Includes $\mu = 68$?
1	$69.407 - 66.692 = 2.715$	Yes
2	$69.426 - 66.490 = 2.936$	Yes
3	$69.211 - 66.741 = 2.470$	Yes
4	$68.050 - 65.766 = 2.284$	Yes

The intervals differ in length, and all four capture $\mu = 68$. If many additional samples of size 20 were generated from this distribution, we would expect about 95% of the confidence intervals created from these samples to capture the number 68. We would expect about 5% of the intervals to be entirely above 68 or entirely below 68.

21. **(a)** $\bar{x} = 25.176 \approx 25.2,$ as stated

        $s = 15.472 \approx 15.5,$ as stated

    (b) $n = 51,$ so $d.f. = n - 1 = 50.$

        $c = 90\%,\ t_{0.90}$ with 50 $d.f. = 1.676$

$$E = \frac{t_c s}{\sqrt{n}} = \frac{1.676(15.5)}{\sqrt{51}} = 3.6$$
$$(\bar{x} - E) < \mu < (\bar{x} + E)$$
$$(25.2 - 3.6) < \mu < (25.2 + 3.6)$$
$$21.6 < \mu < 28.8$$

    (c) $c = 99\%,\ t_{0.99}$ with 50 $d.f. = 2.678$

$$E = \frac{t_c s}{\sqrt{n}} = \frac{2.678(15.5)}{\sqrt{51}} \approx 5.8$$
$$(\bar{x} - E) < \mu < (\bar{x} + E)$$
$$(25.2 - 5.8) < \mu < (25.2 + 5.8)$$
$$19.4 < \mu < 31.0$$

    (d) Using both confidence intervals, we can say that the P/E for Bank One is well below the population average. The P/E for AT&T Wireless is well above the population average. The P/E for Disney is within both confidence intervals. It appears that the P/E for Disney is close to the population average P/E.

    (e) By the central limit theorem, when n is large, the $\bar{x}$ distribution is approximately normal. In general, $n \geq 30$ is considered large.

23. (a) $n = 31,\ d.f. = n - 1 = 30,\ \bar{x} = 45.2,\ s = 5.3$

$c = 90\%,\ t_{0.90} = 1.697,\ E = \dfrac{t_c s}{\sqrt{n}} = \dfrac{1.697(5.3)}{\sqrt{31}} \approx 1.62$

$c = 95\%,\ t_{0.95} = 2.042,\ E = \dfrac{t_c s}{\sqrt{n}} = \dfrac{2.042(5.3)}{\sqrt{31}} \approx 1.94$

$c = 99\%,\ t_{0.99} = 2.750,\ E = \dfrac{t_c s}{\sqrt{n}} = \dfrac{2.750(5.3)}{\sqrt{31}} \approx 2.62$

90% C.I.: $(45.2 - 1.62) < \mu < (45.2 + 1.62)$

$43.58 < \mu < 46.82$

95% C.I.: $(45.2 - 1.94) < \mu < (45.2 + 1.94)$

$43.26 < \mu < 47.14$

99% C.I.: $(45.2 - 2.62) < \mu < (45.2 + 2.62)$

$42.58 < \mu < 47.82$

(b) $n = 31,\ \bar{x} = 45.2,\ \sigma = 5.3$

$c = 90\%,\ z_{0.90} = 1.645,\ E = \dfrac{z_c \sigma}{\sqrt{n}} = \dfrac{1.645(5.3)}{\sqrt{31}} \approx 1.57$

$c = 95\%,\ z_{0.95} = 1.96,\ E = \dfrac{z_c \sigma}{\sqrt{n}} = \dfrac{1.96(5.3)}{\sqrt{31}} \approx 1.87$

$c = 99\%,\ z_{0.99} = 2.58,\ E = \dfrac{z_c \sigma}{\sqrt{n}} = \dfrac{2.58(5.3)}{\sqrt{31}} \approx 2.46$

90% C.I.: $(45.2 - 1.57) < \mu < (45.2 + 1.57)$

$43.63 < \mu < 46.77$

95% C.I.: $(45.2 - 1.87) < \mu < (45.2 + 1.87)$

$43.33 < \mu < 47.07$

99% C.I.: $(45.2 - 2.46) < \mu < (45.2 + 2.46)$

$42.74 < \mu < 47.66$

(c) Yes; the respective intervals based on the student's t distribution are slightly longer.

(d) For student's t, $n = 81,\ d.f. = n - 1 = 80,\ \bar{x} = 45.2,\ s = 5.3$.

$c = 90\%,\ t_{0.90} = 1.664,\ E = \dfrac{t_c s}{\sqrt{n}} = \dfrac{1.664(5.3)}{\sqrt{81}} \approx 0.98$

$c = 95\%,\ t_{0.95} = 1.990,\ E = \dfrac{t_c s}{\sqrt{n}} = \dfrac{1.990(5.3)}{\sqrt{81}} \approx 1.17$

$c = 99\%,\ t_{0.99} = 2.639,\ E = \dfrac{t_c s}{\sqrt{n}} = \dfrac{2.639(5.3)}{\sqrt{81}} \approx 1.55$

90% C.I.: $(45.2 - 0.98) < \mu < (45.2 + 0.98)$

$44.22 < \mu < 46.18$

95% C.I.: $(45.2 - 1.17) < \mu < (45.2 + 1.17)$

$44.03 < \mu < 46.37$

99% C.I.: $(45.2 - 1.55) < \mu < (45.2 + 1.55)$

$43.65 < \mu < 46.75$

For standard normal, $n = 81$, $\bar{x} = 45.2$, $\sigma = 5.3$.

$c = 90\%$, $z_{0.90} = 1.645$, $E = \dfrac{z_c \sigma}{\sqrt{n}} = \dfrac{1.645(5.3)}{\sqrt{81}} \approx 0.97$

$c = 95\%$, $z_{0.95} = 1.96$, $E = \dfrac{z_c \sigma}{\sqrt{n}} = \dfrac{1.96(5.3)}{\sqrt{81}} \approx 1.15$

$c = 99\%$, $z_{0.99} = 2.58$, $E = \dfrac{z_c \sigma}{\sqrt{n}} = \dfrac{2.58(5.3)}{\sqrt{81}} \approx 1.52$

90% C.I.: $(45.2 - 0.97) < \mu < (45.2 + 0.97)$

$\qquad\qquad 44.23 < \mu < 46.17$

95% C.I.: $(45.2 - 1.15) < \mu < (45.2 + 1.15)$

$\qquad\qquad 44.05 < \mu < 46.35$

99% C.I.: $(45.2 - 1.52) < \mu < (45.2 + 1.52)$

$\qquad\qquad 43.68 < \mu < 46.72$

The intervals using the t distribution are still slightly longer than corresponding intervals using the standard normal distribution. However, with a larger sample size, the differences between the two methods are less pronounced.

Section 7.3

Answers may vary slightly owing to rounding.

1. Use $\hat{p} = \dfrac{r}{n}$.

3. **(a)** No, the margin of error only incorporates the sample size, sample proportion, and confidence level.
    (b) It reflects the difference between p and $\hat{p}$. In words, the margin of error is the difference between the population proportion and the sample proportion.

5. No, Jerry does not have a random sample of laptop computers. Also, because all the laptops he tested for spyware are those of students in the same computer class, it could be that students shared software with classmates and spread the infection among the laptops owned by the students of the class.

7. **(a)** Yes. Both $n\hat{p} = 100(0.30) = 30 > 5$ and $n\hat{q} = 100(0.70) = 70 > 5$ and we have binomial trials.
    (b) $\hat{p} - E < p < \hat{p} + E$

$$\hat{p} - z_c \sqrt{\frac{\hat{p}\hat{q}}{n}} < p < \hat{p} + z_c \sqrt{\frac{\hat{p}\hat{q}}{n}}$$

$$0.30 - 1.645 \sqrt{\frac{0.30(0.70)}{100}} < p < 0.30 + 1.645 \sqrt{\frac{0.30(0.70)}{100}}$$

$$0.30 - 0.075 < p < 0.30 + 0.075$$

$$0.225 < p < 0.375$$

   (c) We are 90% confident that this interval is one of the intervals that contains p.

9.   **(a)** $n = p(1-p)\left(\dfrac{z_c}{E}\right)^2 = 0.25(0.75)\left(\dfrac{1.96}{0.1}\right)^2 = 72.03 \rightarrow n = 73$

      (b) $n = p(1-p)\left(\dfrac{z_c}{E}\right)^2 = 0.50(0.50)\left(\dfrac{1.96}{0.1}\right)^2 = 96.04 \rightarrow n = 97$

11. $r = 39,\ n = 62,\ \hat{p} = \dfrac{r}{n} = \dfrac{39}{62},\ \hat{q} = 1 - \hat{p} = \dfrac{23}{62}$

      (a) $\hat{p} = \dfrac{39}{62} = 0.6290$

      (b) $c = 95\%, z_c = z_{0.95} = 1.96$

$$E \approx z_c \sqrt{\hat{p}\hat{q}/n} = 1.96\sqrt{(0.6290)(1-0.6290)/62} = 0.1202$$
$$(\hat{p} - E) < p < (\hat{p} + E)$$
$$(0.6290 - 0.1202) < p < (0.6290 + 0.1202)$$
$$0.5088 < p < 0.7492 \text{ or approximately } 0.51 \text{ to } 0.75.$$

We are 95% confident that the true proportion of actors who are extroverts is between 0.51 and 0.75. In repeated sampling from the same population, approximately 95% of the samples would generate confidence intervals that would capture the true value of $\hat{p}$.

      (c) $np \approx n\hat{p} = r = 39$

$$nq \approx n\hat{q} = n - r = 62 - 39 = 23$$

It is quite likely that np and $nq > 5$ because their estimates are much larger than 5. If np and $nq > 5$, then $\hat{p}$ is approximately normal with $\mu = p$ and $\sigma = \sqrt{pq/n}$.

13. $n = 5222,\ r = 1619$

      (a) $\hat{p} = \dfrac{r}{n} = \dfrac{1619}{5222} = 0.3100$, so $\hat{q} = 1 - \hat{p} = 0.6900$.

      (b) $c = 99\%$, so $z_c = 2.58$

$$E \approx z_c \sqrt{\hat{p}\hat{q}/n} = 2.58\sqrt{(0.3100)(0.6900)/5222} = 0.0165$$
$$(\hat{p} - E) < p < (\hat{p} + E)$$
$$(0.3100 - 0.0165) < p < (0.3100 + 0.0165)$$
$$0.2935 < p < 0.3265 \text{ or approximately } 0.29 \text{ to } 0.33.$$

In repeated sampling, approximately 99% of the confidence intervals generated from the samples would include p, the proportion of judges who are hogans.

      (c) $np \approx n\hat{p} = r = 1619,\ nq \approx n\hat{q} = n(1 - \hat{p}) = n - r = 3603$

Since the estimates of np and nq are much greater than 5, it is reasonable to assume np and $nq > 5$. Then we can use the normal distribution with $\mu = p$ and $\sigma = \sqrt{pq/n}$ to approximate the distribution of $\hat{p}$.

15. $n = 5{,}792$, $r = 3{,}139$

   (a) $\hat{p} = \dfrac{r}{n} = \dfrac{3{,}139}{5{,}792} = 0.5420$, so $\hat{q} = 1 - \hat{p} = 0.4580$.

   (b) $c = 99\%$, so $z_c = 2.58$

   $$E \approx z_c \sqrt{\hat{p}\hat{q}/n} = 2.58\sqrt{(0.5420)(0.4580)/5792} = 0.0169$$
   $$(\hat{p} - E) < p < (\hat{p} + E)$$
   $$(0.5420 - 0.0169) < p < (0.5420 + 0.0169)$$
   $$0.5251 < p < 0.5589, \text{ or approximately } 0.53 \text{ to } 0.56.$$

   If we drew many samples of size 5,792 physicians from those in Colorado and generated a confidence interval from each sample, we would expect approximately 99% of the intervals to include the true proportion of Colorado physicians providing at least some charity care.

   (c) $np \approx n\hat{p} = r = 3139 > 5$; $nq \approx n\hat{q} = n - r = 2653 > 5$.

   Since the estimates of np and nq are much larger than 5, it is reasonable to assume np and nq are both greater than 5. Under the circumstances, it is appropriate to approximate the distribution of $\hat{p}$ with a normal distribution with $\mu = p$ and $\sigma = \sqrt{pq/n}$.

17. $n = 855$, $r = 26$

   (a) $\hat{p} = \dfrac{r}{n} = \dfrac{26}{855} = 0.0304$, so $\hat{q} = 1 - \hat{p} = 0.9696$.

   (b) $c = 99\%$, so $z_c = 2.58$

   $$E \approx z_c \sqrt{\hat{p}\hat{q}/n} = 2.58\sqrt{(0.0304)(0.9696)/855} = 0.0151$$
   $$(\hat{p} - E) < p < (\hat{p} + E)$$
   $$(0.0304 - 0.0151) < p < (0.0304 + 0.0151)$$
   $$0.0153 < p < 0.0455, \text{ or approximately } 0.02 \text{ to } 0.05.$$

   If many additional samples of size $n = 855$ were drawn from this fish population and a confidence interval were created from each such sample, approximately 99% of those confidence intervals would contain p, the catch-and-release mortality rate (barbless hooks removed).

   (c) $np \approx n\hat{p} = r = 26 > 5$; $nq \approx n\hat{q} = n - r = 829 > 5$.

   Based on the estimates of np and nq, it is safe to assume both np and $nq > 5$. When np and $nq > 5$, the distribution of $\hat{p}$ can be approximated accurately by a normal distribution with $\mu = p$ and $\sigma = \sqrt{pq/n}$.

19. $n = 730$, $r = 628$, $n - r = 102$; both np and $nq > 5$.

   (a) $\hat{p} = \dfrac{r}{n} = \dfrac{628}{730} = 0.8603$, so $\hat{q} = 0.1397$.

   (b) $c = 95\%$, so $z_c = 1.96$

   $$E \approx z_c \sqrt{\hat{p}\hat{q}/n} = 1.96\sqrt{(0.8603)(0.1397)/730} = 0.0251$$
   $$(\hat{p} - E) < p < (\hat{p} + E)$$
   $$(0.8603 - 0.0251) < p < (0.8603 + 0.0251)$$
   $$0.8352 < p < 0.8854, \text{ or about } 0.84 \text{ to } 0.89.$$

   In repeated sampling, approximately 95% of the intervals created from the samples would include p, the proportion of loyal women shoppers.

(c)   Margin of error = $E \approx 2.5\%$
A recent study shows that about 86% of women shoppers remained loyal to their favorite supermarket last year. The study's margin of error is 2.5 percentage points.

21. $n = 1{,}000$, $r = 250$, $n - r = 750$; both np and $nq > 5$.

(a) $\hat{p} = \dfrac{r}{n} = \dfrac{250}{1{,}000} = 0.2500$, so $\hat{q} = 0.7500$.

(b) $c = 95\%$, so $z_c = 1.96$

$$E \approx z_c \sqrt{\hat{p}\hat{q}/n} = 1.96\sqrt{(0.2500)(0.7500)/1{,}000} = 0.0268$$
$$(\hat{p} - E) < p < (\hat{p} + E)$$
$$(0.2500 - 0.0268) < p < (0.2500 + 0.0268)$$
$$0.2232 < p < 0.2768, \text{ or about } 0.22 \text{ to } 0.28.$$

(c)   Margin of error = $E \approx 2.7\%$
In a survey of 1,000 large corporations, 25% admitted that, given a choice between equally qualified applicants, they would offer the job to the nonsmoker. The survey's margin of error was 2.7 percentage points.

23.   (a)   Here, 85% sure implies $z_c = 1.44$, and we use $p = 0.50$.

$$n = p(1-p)\left(\frac{z_c}{E}\right)^2 = 0.5 \times 0.5 \times \left(\frac{1.44}{0.05}\right)^2 = 207.36, \text{ so we use } n = 208.$$

(b)   We have $\hat{p} = \dfrac{8}{90} = 0.089$, so the calculation is adjusted as follows:

$$n = p(1-p)\left(\frac{z_c}{E}\right)^2 = 0.089 \times 0.911 \times \left(\frac{1.44}{0.05}\right)^2 = 67.25, \text{ so we use } n = 68.$$

25.   (a)   Here, 99% sure implies $z_c = 2.576$, and we use $p = 0.50$.

$$n = p(1-p)\left(\frac{z_c}{E}\right)^2 = 0.50 \times 0.50 \times \left(\frac{2.576}{0.05}\right)^2 = 663.58, \text{ so we use } n = 664.$$

(b)   We have $\hat{p} = 0.54$, so the calculation is adjusted as follows:

$$n = p(1-p)\left(\frac{z_c}{E}\right)^2 = 0.54 \times 0.46 \times \left(\frac{2.576}{0.05}\right)^2 = 659.33, \text{ so we use } n = 660.$$

27.   (a) $\dfrac{1}{4} - \left(p - \dfrac{1}{2}\right)^2 = \dfrac{1}{4} - \left(p^2 - p + \dfrac{1}{4}\right) = p - p^2 = p(1-p)$

(b)   Since $\left(p - \dfrac{1}{2}\right)^2 \geq 0$, then $\dfrac{1}{4} - \left(p - \dfrac{1}{2}\right)^2 \leq \dfrac{1}{4}$ because we are subtracting a nonnegative value from ¼.

Section 7.4

1.   Two samples are independent if sample data drawn from one population are completely unrelated to the selection of data from the other population.

3.   Josh's will be shorter because the critical value t_c is smaller based on the larger degrees of freedom. Kendra's is more conservative because her value of t_c is larger, resulting in a larger margin of error.

5. We can conclude that $\mu_1 < \mu_2$.

7. **(a)** The normal distribution, by Theorem 7.1, and the fact that the samples are independent and the population standard deviations are known.

(b) $(\bar{x}_1 - \bar{x}_2) - E < \mu_1 - \mu_2 < (\bar{x}_1 - \bar{x}_2) + E;$ $E = z_c\sqrt{\dfrac{\sigma_1^2}{n_1} + \dfrac{\sigma_2^2}{n_2}} = 1.645\sqrt{\dfrac{3^2}{20} + \dfrac{4^2}{25}} \approx 1.717$

$$(12 - 14) - 1.717 < \mu_1 - \mu_2 < (12 - 14) + 1.717$$
$$-3.717 < \mu_1 - \mu_2 < -0.283$$

(c) Now use a Student's t distribution with $df = 20 - 1 = 19$ based on the fact that the original distributions are normal and the population standard deviations are unknown.

(d) $(\bar{x}_1 - \bar{x}_2) - E < \mu_1 - \mu_2 < (\bar{x}_1 - \bar{x}_2) + E;$ $E = t_c\sqrt{\dfrac{s_1^2}{n_1} + \dfrac{s_2^2}{n_2}} = 1.729\sqrt{\dfrac{3^2}{20} + \dfrac{4^2}{25}} \approx 1.805$

$$(12 - 14) - 1.805 < \mu_1 - \mu_2 < (12 - 14) + 1.805$$
$$-3.805 < \mu_1 - \mu_2 < -0.195$$

(e) TI-84 2-SampTInt: $df \approx 42.85;\ (-3.755, -0.2448)$

(f) Since the 90% confidence interval contains all negative values, you can be 90% confident that μ_1 is less than μ_2.

9. **(a)** Yes. $n\hat{p}_1 = 40\left(\dfrac{10}{40}\right) = 10 > 5$, $n\hat{q}_1 = 40\left(\dfrac{30}{40}\right) = 30 > 5$, $n\hat{p}_2 = 50\left(\dfrac{15}{50}\right) = 15 > 5$,

$n\hat{q}_2 = 50\left(\dfrac{35}{50}\right) = 35 > 5$.

(b) $E = z_c\sqrt{\dfrac{\hat{p}_1\hat{q}_1}{n_1} + \dfrac{\hat{p}_2\hat{q}_2}{n_2}} = 1.645\sqrt{\dfrac{0.25(0.75)}{40} + \dfrac{0.30(0.70)}{50}} \approx 0.155$

$$(\hat{p}_1 - \hat{p}_2) - E < p_1 - p_2 < (\hat{p}_1 - \hat{p}_2) + E$$
$$(0.25 - 0.30) - 0.155 < p_1 - p_2 < (0.25 - 0.30) + 0.155$$
$$-0.05 - 0.155 < p_1 - p_2 < -0.05 + 0.155$$
$$-0.205 < p_1 - p_2 < 0.105$$

(c) No. The 90% interval contains both positive and negative values.

11. **(a)** Using a calculator, the means and standard deviations round to the values given.

(b) $c = 90\%$, $d.f. = 12 - 1 = 11$ (smaller of $n_1 - 1$ and $n_2 - 1$), $t_c = 1.796$

$$E = t_c\sqrt{\dfrac{s_1^2}{n_1} - \dfrac{s_2^2}{n_2}} = 1.796\sqrt{\dfrac{170.4^2}{12} + \dfrac{212.1^2}{16}} \approx 129.9$$

$$[(\bar{x}_1 - \bar{x}_2) - E] < (\mu_1 - \mu_2) < [(\bar{x}_1 - \bar{x}_2) + E]$$
$$(747.5 - 738.9) - 129.9 < (\mu_1 - \mu_2) < (747.5 - 738.9) + 129.9$$
$$-121.3 < (\mu_1 - \mu_2) < 138.5$$

(c) Because the interval contains both positive and negative numbers, we cannot say that one region is more interesting than the other at the 90% confidence level.

(d) Student's t because σ_1 and σ_2 are unknown.

13. (a) Using a calculator, the means and standard deviations round to the values given.

(b) $c = 85\%, \, d.f. \approx 16 - 1 = 15$ (smaller of $n_1 - 1$ and $n_2 - 1$), $\, t_c = 1.517$

$$E = t_c \sqrt{\frac{s_1^2}{n_1} + \frac{s_2^2}{n_2}} = 1.517 \sqrt{\frac{7.93^2}{16} + \frac{12.26^2}{17}} \approx 5.42$$

$$[(\bar{x}_2 - \bar{x}_2) - E] < (\mu_1 - \mu_2) < [(\bar{x}_1 - \bar{x}_2) + E]$$

$$[(51.66 - 33.60) - 5.42] < (\mu_1 - \mu_2) < [(51.66 - 33.60) + 5.42]$$

$$12.64\% < (\mu_1 - \mu_2) < 23.48\%$$

(c) Because the interval contains only positive values, we can say at the 85% confidence level that technology companies receive a higher population mean percent of foreign revenue.

(d) Use Student's t because σ_1 and σ_2 are unknown.

15. (a) Using a calculator, the means and standard deviations round to the values given.

(b) $c = 90\%, \, d.f. \approx 40 - 1 = 39$ (smaller of $n_1 - 1$ and $n_2 - 1$)

To use Table 6 in Appendix II, round down to $d.f. = 35, \, t_c = 1.690$.

$$E = t_c \sqrt{\frac{s_1^2}{n_1} + \frac{s_2^2}{n_2}} = 1.690 \sqrt{\frac{0.366^2}{45} + \frac{0.314^2}{40}} \approx 0.125$$

$$[(\bar{x}_1 - \bar{x}_2) - E] < (\mu_1 - \mu_2) < [(\bar{x}_1 - \bar{x}_2) + E]$$

$$[(6.179 - 6.453) - 0.125] < (\mu_1 - \mu_2) < [(6.179 - 6.453) + 0.125]$$

$$-0.399 \text{ feet} < (\mu_1 - \mu_2) < -0.149 \text{ feet}$$

(c) Since the interval contains all negative numbers, it seems that the population mean height of pro football players is less than that of pro basketball players at the 90% confidence level.

(d) Use Student's t because σ_1 and σ_2 are not known. Both samples are large, so no assumptions about the original distribution are needed.

17.

	Sample 1	Sample 2
n	375	571
r	289	23
$\hat{p}$	$289/375 = 0.7707$	$23/571 = 0.0403$

(a) Yes. $n_1\hat{p}_1 = 289, n_1\hat{q}_1 = 86, n_2\hat{p}_2 = 23, n_2\hat{q}_2 = 548$. All four of these estimates are > 5.

(b) $c = 99\%, z_c = 2.58$

$$E \approx z_c \sqrt{\frac{\hat{p}_1\hat{q}_1}{n_1} + \frac{\hat{p}_2\hat{q}_2}{n_2}} = 2.58 \sqrt{\frac{(0.7707)(0.2293)}{375} + \frac{(0.0403)(0.9597)}{571}} = 2.58(0.0232) = 0.0599 \approx 0.06$$

$$[(\hat{p}_1 - \hat{p}_2) - E] < (p_1 - p_2) < [(\hat{p}_1 - \hat{p}_2) + E]$$

$$[(0.7707 - 0.0403) - 0.0599] < (p_1 - p_2) < [(0.7707 - 0.0403) + 0.0599]$$

$$0.6705 < (p_1 - p_2) < 0.7903, \text{ or approximately } 0.67 \text{ to } 0.79$$

(c) Because the confidence interval contains only positive values, $p_1 > p_2$ at the 99% level.

19.

	Sample 1	Sample 2
n	9,340	25,111
$\bar{x}$	63.3	72.1
σ	9.17	12.67

(a) Use the normal distribution since both sample sizes are sufficiently large and both population standard deviations are known.

(b) $c = 99\%, z_c = 2.58$

$$E \approx z_c\sqrt{\frac{\sigma_1^2}{n_1}+\frac{\sigma_2^2}{n_2}} = 2.58\sqrt{\frac{9.17^2}{9,340}+\frac{12.67^2}{25,111}} = 0.3201$$

$$[(\bar{x}_1-\bar{x}_2)-E]<(\mu_1-\mu_2)<[(\bar{x}_1-\bar{x}_2)+E]$$
$$[(63.3-72.1)-0.3201]<(\mu_1-\mu_2)<[(63.3-72.1)+0.3201]$$
$$-9.1201<(\mu_1-\mu_2)<-8.4799, \text{ or about } -9.12 \text{ to } -8.48.$$

(c) The interval includes only negative numbers, leading us to believe that $\mu_1 < \mu_2$ at the 99% confidence level. The mean interval between eruptions during the period 1983–1987 is between 8.48 and 9.12 minutes longer than the mean interval between Old Faithful eruptions during the period 1948–1952.

Comment: It is highly unlikely the data in this problem constitute the required two independent random samples. First, the data are time-series observations and are probably highly correlated. It is possible that the 30-year gap between time periods would be sufficient to wipe out the effects of serial correlation so that the two samples could be considered independent. However, the times within each sample are still correlated, and random samples consist of data that are independent (and identically distributed). Second, the large sample sizes, much larger than needed, might indicate that a census rather than a sample of data was used.

21. (a) Yes, use a normal distribution.

$$n_1 = 210, r_1 = 65, \hat{p}_1 = 65/210 = 0.3095, \hat{q}_1 = \frac{145}{210} = 0.6905$$

$$n_2 = 152, r_2 = 18, \hat{p}_2 = 18/152 = 0.1184, \hat{q}_2 = \frac{134}{152} = 0.8816$$

$n_1\hat{p}_1 = 65, n_1\hat{q}_1 = 145, n_2\hat{p}_2 = 18,$ and $n_2\hat{q}_2 = 134$ are all > 5.

(b) $c = 99\%, z_c = 2.58$

$$E \approx z_c\sqrt{\frac{\hat{p}_1\hat{q}_1}{n_1}+\frac{\hat{p}_2\hat{q}_2}{n_2}} = 2.58\sqrt{\frac{0.3095(0.6905)}{210}+\frac{(0.1184)(0.8816)}{152}} = 2.58(0.0413) = 0.1065$$

$$[(\hat{p}_1-\hat{p}_2)-E]<(p_1-p_2)<[(\hat{p}_1-\hat{p}_2)+E]$$
$$[(0.3095-0.1184)-0.1065]<(p_1-p_2)<[(0.3095-0.1184)+0.1065]$$
$$0.0846<(p_1-p_2)<0.2976, \text{ or about } 0.085 \text{ to } 0.298.$$

(c) The interval consists only of positive values, indicating that $p_1 > p_2$. At the 99% confidence level, the difference in the percentage of traditional Navajo hogans is between 0.085 and 0.298; i.e., there are between 8.5% and 29.8% more hogans in the Fort Defiance region than in the Indian Wells region. If it is true that traditional Navajo tend to live in hogans, then, in terms of percentages, there are more traditional Navajo at Fort Defiance than at Indian Hills.

23. **(a)** Using a calculator, the means and standard deviations round to the values given.

    (b) Use the Student's t distribution because the population standard deviations are unknown and the original distributions are mound-shaped and symmetric

    (c) $c = 85\%$, $d.f. \approx 10 - 1 = 9$ (smaller of $n_1 - 1$ and $n_2 - 1$), $t_c = 1.574$

    $$E = t_c\sqrt{\frac{s_1^2}{n_1} + \frac{s_2^2}{n_2}} = 1.574\sqrt{\frac{8.32^2}{10} + \frac{8.87^2}{18}} \approx 5.3$$

    $$[(\bar{x}_1 - \bar{x}_2) - E] < (\mu_1 - \mu_2) < [(\bar{x}_1 - \bar{x}_2) + E]$$

    $$[(75.80 - 66.83) - 5.3] < (\mu_1 - \mu_2) < [(75.80 - 66.83) + 5.3]$$

    $$3.7 \text{ pounds} < (\mu_1 - \mu_2) < 14.3 \text{ pounds}$$

    (d) Yes, the confidence interval contains values that are all positive, so at the 85% confidence level it appears that the population mean weight of grey wolves from Chihuahua is greater than that of the wolves in Durango.

25. Because the original group of 45 subjects was randomly split into three subgroups of 15, each subgrouping can be considered a random sample, and it is independent of the other subgroups. We will use the Student's t distribution because σ_1 and σ_2 are not known. Since the sample sizes (15) are all less than 30 and the t distribution requires the data to be normal but is robust against some departures from normality, the authors of this study would have to make a case for their self-esteem scores' distributions being at least mound-shaped (unimodal) and symmetric.

    $$c = 85\%, d.f. \approx 15 - 1 = 14, t_c = 1.523$$

    Preliminary calculations:

$\mu_i - \mu_j$	$\bar{x}_i - \bar{x}_j$	$E_{ij} = t_c\sqrt{\dfrac{s_i^2}{n_i} + \dfrac{s_j^2}{n_j}}$
1 versus 2	$19.84 - 19.32 = 0.52$	$1.523\sqrt{\dfrac{3.07^2}{15} + \dfrac{3.62^2}{15}} \approx 1.87$
1 versus 3	$19.84 - 17.88 = 1.96$	$1.523\sqrt{\dfrac{3.07^2}{15} + \dfrac{3.74^2}{15}} \approx 1.90$
2 versus 3	$19.32 - 17.88 = 1.44$	$1.523\sqrt{\dfrac{3.62^2}{15} + \dfrac{3.74^2}{15}} \approx 2.05$

    $$[(\bar{x}_i - \bar{x}_j) - E_{ij}] < (\mu_i - \mu_j) < [(\bar{x}_i - \bar{x}_j) + E_{ij}]$$

    (a) $i = 1, j = 2$:

    $$(0.52 - 1.87) < (\mu_1 - \mu_2) < (0.52 + 1.87)$$

    $$-1.35 < (\mu_1 - \mu_2) < 2.39$$

    (b) $i = 1, j = 3$:

    $$(1.96 - 1.90) < (\mu_1 - \mu_3) < (1.96 + 1.90)$$

    $$0.06 < (\mu_1 - \mu_3) < 3.86$$

    (c) $i = 2, j = 3$:

    $$(1.44 - 2.05) < (\mu_2 - \mu_3) < (1.44 + 2.05)$$

    $$-0.61 < (\mu_2 - \mu_3) < 3.49$$

(d) At the 85% level, we can say that there is no significant difference between the self-esteem scores on competence and social acceptance and no significant difference between self-esteem scores on social acceptance and physical attractiveness. However, the interval estimate for $\mu_1 - \mu_3$ contains only positive numbers, indicating that $\mu_1 > \mu_3$. At the 85% confidence level, we can say that the self-esteem score for competence was between 0.06 and 3.86 points higher than that for physical attractiveness.

Notes: The confidence interval formula used in (a)–(c) is designed to capture the true difference, $\mu_i - \mu_j$, 85% of the time. However, the chance that *all* the intervals will *simultaneously* include the true value is less than 85%. Please reference an advanced text on multiple comparisons. Although (b) showed a *statistically* significant difference (at 85%) between μ_1 and μ_3, the paper's authors would have to argue whether a difference of 0.06 to 3.86 points has any *practical* significance. Statistical significance does not necessarily mean that the results have practical significance.

27. (a) If a 95% interval covers 0, then any confidence interval with a larger confidence level will be wider and consequently also cover 0. Thus a 99% interval also will cover 0. A 90% interval may or may not cover 0 depending on the change to the margin of error when z_c or t_c is changed.

(b) If a 95% interval contains only positive numbers, then a 99% interval, which will be wider, may or may not cover only positive numbers. The increased width may cause the interval to cover 0. On the other hand, a 90% interval, which will be shorter, will cover only positive numbers.

29. $E = z_c \sqrt{\dfrac{p_1 q_1}{n_1} + \dfrac{p_2 q_2}{n_2}} = \dfrac{z_c}{\sqrt{n}} \sqrt{p_1 q_1 + p_2 q_2}$ if $n = n_1 = n_2$.

So $\sqrt{n} E = z_c \sqrt{p_1 q_1 + p_2 q_2}$ Multiply both sides by $\sqrt{n}$.

$\sqrt{n} = \dfrac{z_c}{E} \sqrt{p_1 q_1 + p_2 q_2}$ Divide both sides by E.

$n = \left(\dfrac{z_c}{E}\right)^2 (p_1 q_1 + p_2 q_2)$ Square both sides.

So $n \approx \left(\dfrac{z_c}{E}\right)^2 (\hat{p}_1 \hat{q}_1 + \hat{p}_2 \hat{q}_2)$.

If we have no estimate for p_i, we would use the "worst case" estimate; i.e., the conservative approach would be to make n as large as possible by using $\hat{p}_1 = 0.5$, $\hat{q}_1 = 1 - \hat{p}_1 = 0.5$.

So, in this case,

$n \approx \left(\dfrac{z_c}{E}\right)^2 [(0.5)(0.5) + (0.5)(0.5)] = 0.5 \left(\dfrac{z_c}{E}\right)^2$ or $\left(\dfrac{1}{2}\right)\left(\dfrac{z_c}{E}\right)^2$

Again, all sample size estimates are the minimum number meeting the stated criteria.

(a) $c = 99\%, z_c = 2.58, E = 0.04$

$\hat{p}_1 = \dfrac{289}{375} = 0.7707, \hat{q}_1 = 1 - \hat{p}_1 = 0.2293$

$\hat{p}_2 = \dfrac{23}{571} = 0.0403, \hat{q}_2 = 0.9597$

(Recall that the $n_i p_i$ and $n_i q_i$ conditions were checked in Problem 17.)

$n \approx \left(\dfrac{z_c}{E}\right)^2 (\hat{p}_1 \hat{q}_1 + \hat{p}_2 \hat{q}_2) = \left(\dfrac{2.58}{0.04}\right)^2 [0.7707(0.2293) + 0.0403(0.9597)] = 896.107 \approx 897$

n_1 and n_2 should each be 897 (married couples).

(b) Let $\hat{p}_i = \hat{q}_i = 0.5; c = 95\%, z_c = 1.96, E = 0.05$.

$$n \approx \left(\frac{1}{2}\right)\left(\frac{z_c}{E}\right)^2 = \left(\frac{1}{2}\right)\left(\frac{1.96}{0.05}\right)^2 = 768.32 \approx 769$$

n_1 and n_2 should each be 769 (married couples).

31. (a) $c = 85\%$, $n_1 + n_2 - 2 = 10 + 18 - 2 = 26$, $t_{0.85}$ with 26 $d.f.$ = 1.483

$$s_{\text{pooled}} = \sqrt{\frac{(n_1 - 1)s_1^2 + (n_2 - 1)s_2^2}{n_1 + n_2 - 2}} = \sqrt{\frac{9(8.32^2) + 17(8.87^2)}{10 + 18 - 2}} = 8.6836$$

$$E \approx t_c s_p \sqrt{\frac{1}{n_1} + \frac{1}{n_2}} = 1.483(8.6836)\sqrt{\frac{1}{10} + \frac{1}{18}} = 5.0791$$

$$[(\bar{x}_1 - \bar{x}_2) - E] < (\mu_1 - \mu_2) < [(\bar{x}_1 - \bar{x}_2) + E]$$
$$[(75.80 - 66.83) - 5.0791] < (\mu_1 - \mu_2) < [75.80 - 66.83) + 5.0791]$$
$$3.9 < (\mu_1 - \mu_2) < 14.1$$

(b) The pooled standard deviation method has a shorter interval and larger degrees of freedom.

Chapter Review Problems

1.   point estimate: A single number used to estimate a population parameter, such as $\bar{x}$ for μ or $\hat{p}$ for p.

critical value: The x-axis values of a probability density function, such as the standard normal or Student's t, which cut off an area of c, $0 \le c \le 1$, under the curve between them. *Examples:* The area under the standard normal curve between $-z_c$ and $+z_c$ is c; the area under the curve of a Student's t distribution between $-t_c$ and $+t_c$ is c. The area is symmetric about the curve's mean μ.

maximal margin of error E: The largest distance ("error") between the point estimate and the parameter it estimates that can be tolerated under certain circumstances; E is the half-width of a confidence interval.

confidence level c: A measure of the reliability of an interval estimate: c denotes the proportion of all possible confidence interval estimates of a parameter (or difference between two parameters) that will capture the true value being estimated. It is a statement about the probability the *procedure* being used has of capturing the value of interest. It *cannot* be considered a measure of the reliability of a *specific* interval because any specific interval is either right or wrong—either it captures the parameter value or it does not.

confidence interval: A procedure designed to give a range of values as an estimate of an unknown parameter value. A 90% confidence interval for μ says that if all possible samples of size n were drawn, and a 90% confidence interval for μ was created for each such sample using the prescribed method, then if the true value of μ were known, 90% of the confidence intervals created would capture the value of μ.

3.   (a) No. The probability that μ falls inside the interval is 0 or 1.
(b) Yes. By definition, 99% confidence intervals are constructed such that, over the long run, 99% of all intervals will capture the population mean μ.

5. $n = 73,\ \bar{x} = 178.70,\ \sigma = 7.81,\ c = 95\%,\ z_c = 1.96$

$$E \approx \frac{z_c \sigma}{\sqrt{n}} = \frac{1.96(7.81)}{\sqrt{73}} = 1.7916 \approx 1.79$$

$$(\bar{x} - E) < \mu < (\bar{x} + E)$$
$$(178.70 - 1.79) < \mu < (178.70 + 1.79)$$
$$176.91 < \mu < 180.49$$

7. **(a)** $\bar{x} = 74.2,\ s = 18.2530 \approx 18.3$, as indicated.

   (b) $c = 95\%, n = 15,\ t_{0.95}$ with $n - 1 = 14\ d.f. = 2.145$.

   $$E \approx \frac{t_c s}{\sqrt{n}} = \frac{2.145(18.3)}{\sqrt{15}} = 10.1352 \approx 10.1$$

   $$(\bar{x} - E) < \mu < (\bar{x} + E)$$
   $$(74.2 - 10.1) < \mu < (74.2 + 10.1)$$
   $$64.1\ \text{cm} < \mu < 84.3\ \text{cm}$$

9. $n = 2958, r = 1538, \hat{p} = \dfrac{r}{n} = \dfrac{1538}{2958} = 0.5199 \approx 0.52$

   $\hat{q} = 1 - \hat{p} = 0.4801, n - r = 1420, c = 90\%, z_c = 1.645$

   $np \approx n\hat{p} = r = 1{,}538 > 5, nq \approx n\hat{q} = n - r = 1{,}420 > 5$

   $$E \approx z_c \sqrt{\frac{\hat{p}\hat{q}}{n}} = 1.645 \sqrt{\frac{(0.5199)(0.4801)}{2{,}958}} = 0.0151 \approx 0.02$$

   $$(\hat{p} - E) < p < (\hat{p} + E)$$
   $$(0.52 - 0.02) < p < (0.52 + 0.02)$$
   $$0.50 < p < 0.54$$

11. $n = 167, r = 68$

   (a) $\hat{p} = \dfrac{r}{n} = \dfrac{68}{167} = 0.4072, \hat{q} = 0.5928, n - r = 99$

   (b) $n\hat{p} = r = 68 > 5$ and $n\hat{q} = n - r = 99 > 5$

   $c = 95\%, z_c = 1.96$

   $$E \approx z_c \sqrt{\frac{\hat{p}\hat{q}}{n}} = 1.96 \sqrt{\frac{(0.4072)(0.5928)}{167}} = 0.0745$$

   $$(\hat{p} - E) < p < (\hat{p} + E)$$
   $$(0.4072 - 0.0745) < p < (0.4072 + 0.0745)$$
   $$0.3327 < p < 0.4817, \text{ or about } 0.333 \text{ to } 0.482$$

13. **(a)** Using a calculator, the means and standard deviations round to the values given.

 (b) $c = 95\%$, $d.f.$ is smaller of $n_1 - 1$ and $n_2 - 1$, $d.f. \approx n_1 - 1 = 72 - 1 = 71$ (round down to 70 for Table 6). $t_c = 1.994$

$$E = t_c \sqrt{\frac{s_1^2}{n_1} + \frac{s_2^2}{n_2}} = 1.994 \sqrt{\frac{2.08^2}{72} + \frac{3.03^2}{80}} \approx 0.83$$

$$[(\bar{x}_1 - \bar{x}_2) - E] < (\mu_1 - \mu_2) < [(\bar{x}_1 - \bar{x}_2) + E]$$
$$[(11.42 - 10.65) - 0.83] < (\mu_1 - \mu_2) < [(11.42 - 10.65) + 0.83]$$
$$-0.06 < (\mu_1 - \mu_2) < 1.6$$

 (c) Because the interval contains both positive and negative values, we cannot conclude that there is any difference in soil water content in the two fields at the 95% confidence level.

 (d) Student's t distribution because σ_1 and σ_2 are not known. Both samples are large, so no assumptions about the original distribution are needed.

15. $n_1 = 18, \bar{x}_1 = 98, s_1 = 6.5$

 $n_2 = 24, \bar{x}_2 = 90, s_2 = 7.3$

 (a) $c = 75\%$, $d.f. \approx 18 - 1 = 17$ (smaller of $n_1 - 1$ and $n_2 - 1$); $t_c = 1.191$

$$E = t_c \sqrt{\frac{s_1^2}{n_1} + \frac{s_2^2}{n_2}} = 1.191 \sqrt{\frac{6.5^2}{18} + \frac{7.3^2}{24}} \approx 2.5$$

$$[(\bar{x}_1 - \bar{x}_2) - E] < (\mu_1 - \mu_2) < [(\bar{x}_1 - \bar{x}_2) + E]$$
$$[(98 - 90) - 2.5] < (\mu_1 - \mu_2) < [(98 - 90) + 2.5]$$
$$5.5 \text{ pounds} < (\mu_1 - \mu_2) < 10.5 \text{ pounds}$$

 (b) Since the interval contains only positive values, we can say at the 75% confidence level that $\mu_1 > \mu_2$, i.e., that Canadian wolves weigh more than Alaskan wolves, and that the difference is approximately 5.5–10.5 pounds.

17. $n_1 = 93, r_1 = 79, \hat{p}_1 = \dfrac{79}{93} = 0.8495, \hat{q}_1 = 0.1505, n_1 - r_1 = 14$

 $n_2 = 83, r_2 = 74, \hat{p}_2 = \dfrac{74}{83} = 0.8916, \hat{q}_2 = 0.1084, n_2 - r_2 = 9$

 $n_i \hat{p}_i$ and $n_i \hat{q}_i$ are both > 5.

 (a) $c = 95\%, z_c = 1.96$

$$E \approx z_c \sqrt{\frac{\hat{p}_1 \hat{q}_1}{n_1} + \frac{\hat{p}_2 \hat{q}_2}{n_2}} = 1.96 \sqrt{\frac{(0.8495)(0.1505)}{93} + \frac{(0.8916)(0.1084)}{83}} = 0.0988$$

$$[(\hat{p}_1 - \hat{p}_2) - E] < (p_1 - p_2) < [(\hat{p}_1 - \hat{p}_2) + E]$$
$$[(0.8495 - 0.8916) - 0.0988] < (p_1 - p_2) < [(0.8495 - 0.8916) + 0.0988]$$
$$-0.1409 < (p_1 - p_2) < 0.0567$$

 (b) Since the interval contains positive and negative values, we can say, at the 95% level, that there is no significant difference between the proportion of accurate responses for face-to-face interviews and that for telephone interviews.

19. (a) $P(A_1 < \mu_1 < B_1) = 0.80$

$P(A_2 < \mu_2 < B_2) = 0.80$

That is, the two intervals are designed so that the confidence interval procedure produces intervals A_i to B_i that capture μ_i 80% of the time.

$$P(A_1 < \mu_1 < B_1 \text{ and } A_2 < \mu_2 < B_2) = P(A_1 < \mu_1 < B_1) \times P(A_2 < \mu_2 < B_2 \mid A_1 < \mu_1 < B_1)$$

But the intervals were created using independent samples, so the intervals themselves are independent. Thus

$$P(A_1 < \mu_1 < B_1 \text{ and } A_2 < \mu_2 < B_2) = P(A_1 < \mu_1 < B_1) \times P(A_2 < \mu_2 < B_2)$$

By the definition of independent events,

$$0.80 \times 0.80 = 0.64$$

The probability that both intervals are simultaneously correct, i.e., that both intervals capture μ_i, is 0.64.

$$\begin{aligned} P(\text{at least one interval fails to capture its } \mu_i) &= 1 - P(\text{both intervals capture their } \mu_1) \\ &= 1 - 0.64 \\ &= 0.36 \end{aligned}$$

(b) $P(A_1 < \mu_1 < B_1) = c$

$P(A_2 < \mu_2 < B_2) = c$

(Both confidence intervals are at level c.)

If $0.90 = c^2$, then $\sqrt{0.90} = c = 0.9487$, or about 0.95.

(c) Answers vary. In large, complex engineering designs, each component must be within design specifications or the project will fail.

Consider the hundreds of components that must function properly to launch the space shuttle, keep it orbiting, and return it safely to earth. For example, nuts, bolts, and rivets must be a certain size, give or take some tiny amount. Tiles must be able to withstand a huge range of temperatures, from the ambient air temperature at launch time to the extreme heat of re-entry.

Each of the design specifications can be thought of as a confidence interval. Manufacturers and suppliers want to be very confident that their parts are well within the specifications, or they might lose their contracts to competitors. Similarly, NASA wants to be very confident that all the parts, as a group, meet specifications; otherwise, costly delays or catastrophic failures may occur.

Recall that much of the *Challenger* disaster was due to O-ring failure—because NASA decided to go ahead with the launch even though it had been warned by the O-ring manufacturer that the temperature at the launch site was below the lowest temperature at which the O-rings had been tested and that the O-rings might fail at that temperature.

If NASA will tolerate only a 1 in 1,000 or 1 in 1,000,000 chance of failure, i.e., $c = 0.999$ or $c = 0.999999$, the individual components' confidence levels c must be (much) higher than NASA's.

Chapter 8: Hypothesis Testing

Section 8.1

Note: **For all graphs provided, the *P* value is indicated by the shaded portion in the tails.**

1. See text for definitions. Essays may include
   (a) A working hypothesis about the population parameter in question is called the *null hypothesis*. The value specified in the null hypothesis is often a historical value, a claim, or a production specification.
   (b) Any hypothesis that differs from the null hypothesis is called an *alternate hypothesis*.
   (c) If we reject the null hypothesis when it is in fact true, we have an error that is called a *type I error*. On the other hand, if we fail to reject the null hypothesis when it is in fact false, we have made an error that is called a *type II error*.
   (d) The probability with which we are willing to risk a type I error is called the *level of significance* of a test. The probability of making a type II error is denoted by β.

3. No, if we fail to reject the null hypothesis, we have not proven it to be true beyond all doubt. The evidence is not sufficient to merit rejecting H_0.

5. The probability of rejecting the null hypothesis when it is true is called the level of significance. The symbol is α, also the probability of a type I error.

7. Fail to reject H_0.

9. $P-\text{value} = 2(0.0092) = 0.0184$

11. (a) $H_0 : \mu = 40$ (b) $H_0 : \mu \neq 40$ (c) $H_0 : \mu > 40$ (d) $H_0 : \mu < 40$

13. (a) Yes, because the *x* distribution is normal.
    (b) $z = \dfrac{\bar{x} - \mu}{\sigma / \sqrt{n}} = \dfrac{8 - 7}{4 / \sqrt{20}} \approx 1.12$
    (c) $P-\text{value} = 2P\left(z > |z_{\bar{x}}|\right) = 2P\left(z > |1.12|\right) \approx 0.2627$
    (d) Since $P-\text{value} = 0.2627 > \alpha = 0.05$, do not reject the null hypothesis.

15. (a) The claim is $\mu = 60$ kg, so you would use $H_0 : \mu = 60$ kg.
    (b) We want to know if the average weight is less than 60 kg, so you would use $H_1 : \mu < 60$ kg.
    (c) We want to know if the average weight is greater than 60 kg, so you would use $H_1 : \mu > 60$ kg.
    (d) We want to know if the average weight is different from 60 kg, so you would use $H_1 : \mu \neq 60$ kg.
    (e) Since part (b) is a left-tailed test, the critical region is on the left. Since part (c) is a right-tailed test, the critical region is on the right. Since part (d) is a two-tailed test, the critical region is on both sides of the mean.

17. (a) The claim is $\mu = 16.4$ ft, so $H_0 : \mu = 16.4$ ft.
    (b) You want to know if the average is getting larger, so $H_1 : \mu > 16.4$ ft.
    (c) You want to know if the average is getting smaller, so $H_1 : \mu < 16.4$ ft.
    (d) You want to know if the average is different from 16.4 ft, so $H_1 : \mu \neq 16.4$ ft.

(e)   Since part (b) is a right-tailed test, the area corresponding to the P value is on the right. Since part (c) is a left-tailed test, the area corresponding to the P value is on the left. Since part (d) is a two-tailed test, the area corresponding to the P value is on both sides of the mean.

19. (a) $\alpha = 0.01$

$H_0: \mu = 4.7\%$

$H_1: \mu > 4.7\%$

Since > is in H_1, use a right-tailed test.

(b)   Use the standard normal distribution. We assume x has a normal distribution with known standard deviation σ. Note that μ is given in the null hypothesis.

$$z = \frac{\overline{x} - \mu}{\dfrac{\sigma}{\sqrt{n}}} = \frac{5.38 - 4.7}{\dfrac{2.4}{\sqrt{10}}} \approx 0.90$$

(c) P value $= P(z > 0.90) = 0.1841$

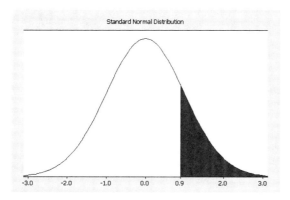

(d)   Since a P value of $0.1841 > 0.01$ for α, we fail to reject H_0. The data are not statistically significant.

(e)   There is insufficient evidence at the 0.01 level to reject the claim that average yield for bank stocks equals average yield for all stocks.

21. (a) $\alpha = 0.01$

$H_0: \mu = 4.55$ g

$H_1: \mu < 4.55$ g

Since < is in H_1, use a left-tailed test.

(b)   Use the standard normal distribution. We assume that x has a normal distribution with known standard deviation σ. Note that μ is given in the null hypothesis.

$$z = \frac{\overline{x} - \mu}{\dfrac{\sigma}{\sqrt{n}}} = \frac{3.75 - 4.55}{\dfrac{0.70}{\sqrt{6}}} \approx -2.80$$

(c) P value $= P(z < -2.80) = 0.0026$

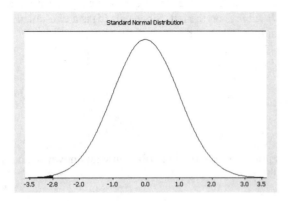

(d) Since a P value of $0.0026 \leq 0.01$, we reject H_0. Yes, the data are statistically significant.

(e) The sample evidence is sufficient at the 0.01 level to justify rejecting H_0. It seems the humming birds in the Grand Canyon region have a lower average weight.

23. (a) $\alpha = 0.01$

$H_0: \mu = 11\%$

$H_1: \mu \neq 11\%$

Since $\neq$ is in H_1, use a two-tailed test.

(b) Use the standard normal distribution. We assume that x has a normal distribution with known standard deviation σ. Note that μ is given in the null hypothesis.

$$z = \frac{\bar{x} - \mu}{\frac{\sigma}{\sqrt{n}}} = \frac{12.5 - 11}{\frac{5.0}{\sqrt{16}}} = 1.20$$

(c) P value $= 2P(z > 1.20) = 2(0.1151) = 0.2302$

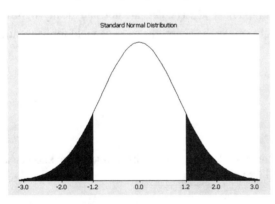

(d) Since a P value of $0.2302 > 0.01$ for α, we fail to reject H_0. The data are not statistically significant.

(e) There is insufficient evidence at the 0.01 level of significance to reject H_0. It seems the average hail damage to wheat crops in Weld County matches the national average.

Section 8.2

1. The P value for a two-tailed test of μ is twice the P value for a one-tailed test based on the same sample data and null hypothesis.

3. Use $n - 1$ degrees of freedom.

5. Yes. If the P value is less than $\alpha = 0.01$, then it also will be less than $\alpha = 0.05$. In both cases, reject the null hypothesis.

7. **(a)** $0.010 < P - \text{value} < 0.020$; technology gives $P - \text{value} \approx 0.015$.
   (b) $0.005 < P - \text{value} < 0.010$; technology gives $P - \text{value} \approx 0.0075$.

9. **(a)** Yes, because the x distribution is mound-shaped and symmetric and σ is unknown; $df = n - 1 = 25 - 1 = 24$

   (b) $H_0 : \mu = 9.5; \quad H_1 : \mu \neq 9.5;$

   (c) $t = \dfrac{\bar{x} - \mu}{s / \sqrt{n}} = \dfrac{10 - 9.5}{2 / \sqrt{25}} = 1.25$

   (d) Using Table 6, $0.200 < P - \text{value} < 0.250$.
   (e) Since $P - \text{value} > \alpha = 0.05$, do not reject the null hypothesis.
   (f) The sample evidence is insufficient at the 0.05 level to reject the null hypothesis.

11. **(a)** $\alpha = 0.01$
   $H_0: \mu = 16.4$ ft
   $H_1: \mu > 16.4$ ft

   (b) Use the standard normal distribution. The sample size is large, $n \geq 30$, and $\sigma = 3.5$.
   $z = \dfrac{\bar{x} - \mu}{\dfrac{\sigma}{\sqrt{n}}} = \dfrac{17.3 - 16.4}{\dfrac{3.5}{\sqrt{36}}} \approx 1.54$

   (c) $P \text{ value} = P(z > 1.54) = 0.0618$

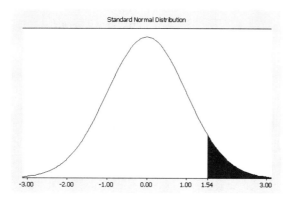

   (d) Since a P value of $0.0618 > 0.01$, we fail to reject H_0. The data are not statistically significant.
   (e) At the 1% level of significance, there is insufficient evidence to say average storm level is increasing.

13. **(a)** $\alpha = 0.01$
   $H_0: \mu = 1.75$ years
   $H_1: \mu > 1.75$ years

(b) Use the Student's t distribution with $d.f. = n - 1 = 46 - 1 = 45$. The sample size is large, $n \geq 30$, and σ is unknown.

$$t = \frac{\bar{x} - \mu}{\frac{s}{\sqrt{n}}} = \frac{2.05 - 1.75}{\frac{0.82}{\sqrt{46}}} \approx 2.481$$

(c) For $d.f. = 45$, 2.481 falls between entries 2.412 and 2.690. Use one-tailed areas to find that $0.005 < P$ value < 0.010. Using a TI-84, P value ≈ 0.0084.

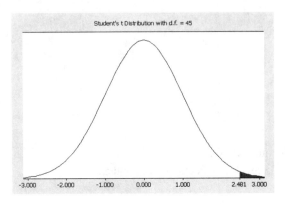

Student's t Distribution with d.f. = 45

(d) Since the entire P-value interval ≤ 0.01, we reject H_0. Yes, the data are statistically significant.

(e) At the 1% level of significance, sample data indicate that the average age of Minnesota region coyotes is greater than 1.75 years.

15. (a) $\alpha = 0.05$

$H_0: \mu = 19.4$

$H_1: \mu \neq 19.4$

(b) Use the Student's t distribution with $d.f. = n - 1 = 36 - 1 = 35$. The sample size is large, $n \geq 30$, and σ is unknown.

$$t = \frac{\bar{x} - \mu}{\frac{s}{\sqrt{n}}} = \frac{17.9 - 19.4}{\frac{5.2}{\sqrt{36}}} \approx -1.731$$

(c) For $d.f. = 35$, 1.731 falls between entries 1.690 and 2.030. Use two-tailed areas to find that $0.05 < P$ value < 0.100. Using a TI-84, P value ≈ 0.0923.

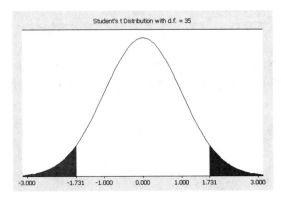

Student's t Distribution with d.f. = 35

(d) Since the P-value interval > 0.05, we fail to reject H_0. The data are not statistically significant.

(e) At the 5% level of significance, the sample evidence does not support rejecting the claim that the average P/E for socially responsible funds is different from that of the S&P stock index.

17. (i) Use a calculator to verify. Rounded values are used in part (ii).
    (ii) (a) $\alpha = 0.05$

    $H_0: \mu = 4.8$

    $H_1: \mu < 4.8$

    (b) Use the Student's t distribution with $d.f. = n - 1 = 6 - 1 = 5$. We assume that x has a distribution that is approximately normal and that σ is unknown.

    $$t = \frac{\bar{x} - \mu}{\frac{s}{\sqrt{n}}} = \frac{4.40 - 4.8}{\frac{0.28}{\sqrt{6}}} \approx -3.499$$

    (c) For $d.f. = 5$, 3.499 falls between entries 3.365 and 4.032. Use one-tailed areas to find that $0.005 < P$ value < 0.010. Using a TI-84, P value ≈ 0.0086.

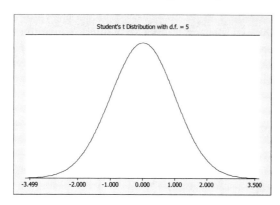

Student's t Distribution with d.f. = 5

    (d) Since the entire P-value interval ≤ 0.05, we reject H_0. Yes, the data are statistically significant.

    (e) At the 5% level of significance, sample evidence supports the claim that the average RBC count for this patient is less than 4.8.

19. (i) Use a calculator to verify. Rounded values are used in part (ii).
    (ii) (a) $\alpha = 0.01$

    $H_0: \mu = 67$

    $H_1: \mu \neq 67$

    (b) Use the Student's t distribution with $d.f. = n - 1 = 16 - 1 = 15$. We assume that x has a distribution that is approximately normal and that σ is unknown.

    $$t = \frac{\bar{x} - \mu}{\frac{s}{\sqrt{n}}} = \frac{61.8 - 67}{\frac{10.6}{\sqrt{16}}} \approx -1.962$$

    (c) For $d.f. = 15$, 1.962 falls between entries 1.753 and 2.131. Use two-tailed areas to find that $0.050 < P$ value < 0.100. Using a TI-84, P value ≈ 0.0686.

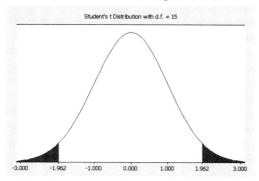

Student's t Distribution with d.f. = 15

(d) Since P-value interval > 0.01, we fail to reject H_0. The data are not statistically significant.

(e) At the 1% level of significance, sample evidence does not support a claim that the average thickness of slab avalanches in Vail is different from that of those in Canada.

21. (i) Use a calculator to verify. Rounded values are used in part (ii).

(ii) (a) $\alpha = 0.05$

$H_0: \mu = 8.8$

$H_1: \mu \neq 8.8$

(b) Use the Student's t distribution with $d.f. = n - 1 = 14 - 1 = 13$. We assume that x has a distribution that is approximately normal and that σ is unknown.

$$t = \frac{\bar{x} - \mu}{\dfrac{s}{\sqrt{n}}} = \frac{7.36 - 8.8}{\dfrac{4.03}{\sqrt{14}}} \approx -1.337$$

(c) For $d.f. = 13$, 1.337 falls between entries 1.204 and 1.350. Use two-tailed areas to find that $0.200 < P$ value < 0.250. Using a TI-84, P value ≈ 0.2042.

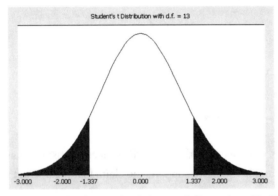

(d) Since P-value interval > 0.05, we fail to reject H_0. No, the data are not statistically significant.

(e) At the 5% level of significance, we cannot conclude that the catch is different from 8.8 fish per day.

23. (a) The P value of a one-tailed test is smaller. For a two-tailed test, the P value is double because it includes the area in both tails.

(b) Yes; the P value of a one-tailed test is smaller, so it might be smaller than α, whereas the P value of a two-tailed test is larger than α.

(c) Yes; if the two-tailed P value is less than α, the one-tail area is also less than α.

(d) Yes, the conclusions can be different. The conclusion based on the two-tailed test is more conservative in the sense that the sample data must be more extreme (differ more from H_0) in order to reject H_0.

25. (a) $H_0: \mu = 20$

$H_1: \mu \neq 20$

For $\alpha = 0.01$, $c = 1 - 0.01 = 0.99$, $\sigma = 4$, and $z_c = 2.58$.

$$E \approx z_c \frac{\sigma}{\sqrt{n}} = 2.58 \frac{4}{\sqrt{36}} = 1.72$$

$$\bar{x} - E < \mu < \bar{x} + E$$

$$22 - 1.72 < \mu < 22 + 1.72$$

$$20.28 < \mu < 23.72$$

The hypothesized mean $\mu = 20$ is not in the interval. Therefore, we reject H_0.

(b) Because $n = 36$ is large, the sampling distribution of $\bar{x}$ is approximately normal by the central limit theorem, and we know σ.

$$z = \frac{\bar{x} - \mu}{\sigma/\sqrt{n}} = \frac{22 - 20}{4/\sqrt{36}} = 3.00$$

From Table 5, P value $= 2P(z < -3.00) = 2(0.0013) = 0.0026$. Since $0.0026 \leq 0.01$, we reject H_0. The results are the same.

27. For a right-tailed test and $\alpha = 0.01$, the critical value is $z_0 = 2.33$; critical region is values to the right of 2.33. Since the sample statistic $z = 1.54$ is not in the critical region, fail to reject H_0. At the 1% level, there is insufficient evidence to say that the average storm level is increasing. Conclusion is the same as with the P-value method.

29. For a right-tailed test and $\alpha = 0.01$, critical value is $t_0 = 2.412$ with $d.f. = 45$. Critical region is values to the right of 2.412. Since the sample test statistic $t = 2.481$ is in the critical region, reject H_0. At the 1% level, sample data indicate that the average age of Minnesota region coyotes is higher than 1.75 years. The conclusion is the same as with the P-value method.

Section 8.3

1. The value of p comes from H_0. Note that $q = 1 - p$.

3. Yes. The corresponding P value for a one-tailed test is half that of a two-tailed test. Thus the P value for the one-tailed test is also less than 0.01.

5. **(a)** Yes, $np = 30(0.5) = 15 > 5$; $nq = 30(0.5) = 15 > 5$.

 (b) $H_0 : p = 0.50$; $H_1 : p \neq 0.50$

 (c) $\hat{p} = \dfrac{12}{30} = 0.4$; $z = \dfrac{\hat{p} - p}{\sqrt{\dfrac{pq}{n}}} = \dfrac{0.4 - 0.5}{\sqrt{\dfrac{0.5(0.5)}{30}}} \approx -1.10$

 (d) $P - \text{value} = 2P\left(z > \left|z_{\hat{p}}\right|\right) = 2P\left(z > \left|-1.10\right|\right) \approx 0.2713$

 (e) Since $P - \text{value} = 0.2713 > \alpha = 0.05$, do not reject the null hypothesis.

 (f) The sample proportion based on 30 trials is not sufficiently different from 0.50 to justify rejecting the null hypothesis at the 5% level of significance.

7. **(i)** **(a)** $\alpha = 0.01$

 $H_0 : p = 0.301$

 $H_1 : p < 0.301$

 (b) Use the standard normal distribution. The $\hat{p}$ distribution is approximately normal when n is sufficiently large, which it is here, because $np = 215(0.301) \approx 647$ and $nq = 215(0.699) \approx 150.3$ are both > 5.

 $\hat{p} = \dfrac{r}{n} = \dfrac{46}{215} \approx 0.214$

 $z = \dfrac{\hat{p} - p}{\sqrt{\dfrac{pq}{n}}} = \dfrac{0.214 - 0.301}{\sqrt{\dfrac{0.301(0.699)}{215}}} \approx -2.78$

(c) P value $= P(z < -2.78) = 0.0027$

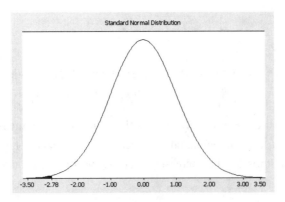

(d) Since a P value of $0.0027 \leq 0.01$, we reject H_0. Yes, the data are statistically significant.

(e) At the 1% level of significance, the sample data indicate that the population proportion of numbers with a leading 1 in the revenue file is less than 0.301 predicted by Benford's law.

(ii) Yes. The revenue data file seems to include more numbers with higher first nonzero digits than Benford's law predicts.

(iii) We have not proved H_0 to be false. However, because our sample data lead us to reject H_0 and conclude that there are too few numbers with a leading digit 1, more investigation is merited.

9. **(a)** $\alpha = 0.01$

$H_0: p = 0.70$

$H_1: p \neq 0.70$

(b) Use the standard normal distribution. The $\hat{p}$ distribution is approximately normal when n is sufficiently large, which it is here, because $np = 32(0.7) = 22.4$ and $nq = 32(0.3) = 9.6$, and both are greater than 5.

$$\hat{p} = \frac{r}{n} = \frac{24}{32} \approx 0.75$$

$$z = \frac{\hat{p} - p}{\sqrt{\dfrac{pq}{n}}} = \frac{0.75 - 0.70}{\sqrt{\dfrac{0.70(0.30)}{32}}} \approx 0.62$$

(c) P value $= 2P(z > 0.62) = 2(0.2676) = 0.5352$

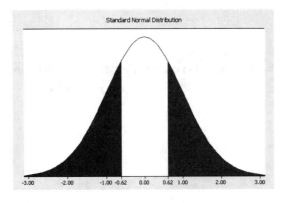

(d) Since a P value of $0.5352 > 0.01$, we fail to reject H_0. No, the data are not statistically significant.

(e) At the 1% level of significance, we cannot say that the population proportion of arrests of males aged 15 to 34 in Rock Springs is different than 70%.

11. (a) $\alpha = 0.01$

    $H_0: p = 0.77$

    $H_1: p < 0.77$

    (b) Use the standard normal distribution. The $\hat{p}$ distribution is approximately normal when n is sufficiently large, which it is here, because $np = 27(0.77) = 20.79$ and $nq = 27(0.23) = 6.21$, and both are greater than 5.

    $$\hat{p} = \frac{r}{n} = \frac{15}{27} = 0.5556$$

    $$z = \frac{\hat{p} - p}{\sqrt{\frac{pq}{n}}} = \frac{0.5556 - 0.77}{\sqrt{\frac{0.77(0.23)}{27}}} = -2.65$$

    (c) P value $= P(z < -2.65) = 0.0004$

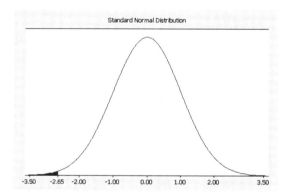

Standard Normal Distribution

    (d) Since a P value of $0.0004 \leq 0.01$, we reject H_0. Yes, the data are statistically significant.

    (e) At the 1% level of significance, the data show that the population proportion of driver fatalities related to alcohol is less than 77% in Kit Carson County.

13. (a) $\alpha = 0.01$

    $H_0: p = 0.50$

    $H_1: p < 0.50$

    (b) Use the standard normal distribution. The $\hat{p}$ distribution is approximately normal when n is sufficiently large, which it is here, because $np = 34(0.50) = 17$ and $nq = 17$, and both are greater than 5.

    $$\hat{p} = \frac{r}{n} = \frac{10}{34} = 0.2941$$

    $$z = \frac{\hat{p} - p}{\sqrt{\frac{pq}{n}}} = \frac{0.2941 - 0.50}{\sqrt{\frac{0.5(0.5)}{34}}} = -2.40$$

    (c) P value $= P(z < -2.40) = 0.0082$

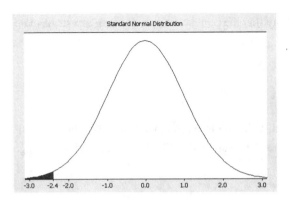

(d) Since a P value of $0.0082 \leq 0.01$, we reject H_0. Yes, the data are statistically significant.

(e) At the 1% level of significance, the data indicate that the population proportion of female wolves is now less than 50% in the region.

15. (a) $\alpha = 0.01$

$H_0: p = 0.261$

$H_1: p \neq 0.261$

(b) Use the standard normal distribution. The $\hat{p}$ distribution is approximately normal when n is sufficiently large, which it is here, because $np = 317(0.261) = 82.737$ and $nq = 317(0.739) = 234.263$, and both are greater than 5.

$$\hat{p} = \frac{r}{n} = \frac{61}{317} = 0.1924$$

$$z = \frac{\hat{p} - p}{\sqrt{\dfrac{pq}{n}}} = \frac{0.1924 - 0.261}{\sqrt{\dfrac{0.261(0.739)}{317}}} \approx -2.78$$

(c) P value $= 2P(z < -2.78) = 2(0.0027) = 0.0054$

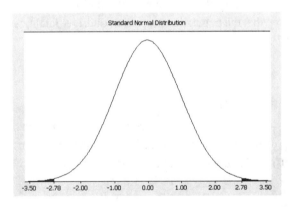

(d) Since a P value of $0.0054 \leq 0.01$, we reject H_0. Yes, the data are statistically significant.

(e) At the 1% level of significance, the sample data indicate that the population proportion of the five-syllable sequence is different from the text of Plato's *Republic*.

17. (a) $\alpha = 0.01$

$H_0: p = 0.47$

$H_1: p > 0.47$

(b) Use the standard normal distribution. The $\hat{p}$ distribution is approximately normal when n is sufficiently large, which it is here, because $np = 1006(0.47) = 472.82$ and $nq = 1006(0.53) = 533.18$, and both are greater than 5.

$$\hat{p} = \frac{r}{n} = \frac{490}{1006} = 0.4871$$

$$z = \frac{\hat{p} - p}{\sqrt{\dfrac{pq}{n}}} = \frac{0.4871 - 0.47}{\sqrt{\dfrac{0.47(0.53)}{1006}}} = 1.09$$

(c) P value $= P(z > 1.09) = 0.1379$

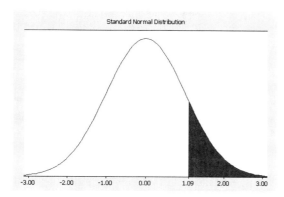

(d) Since a P of $0.1379 > 0.01$, we fail to reject H_0. No, the data are not statistically significant.

(e) At the 1% level of significance, there is insufficient evidence to support the claim that the population proportion of customers loyal to Chevrolet is more than 47%.

19. **(a)** $\alpha = 0.05$

$H_0: p = 0.092$

$H_1: p > 0.092$

(b) Use the standard normal distribution. The $\hat{p}$ distribution is approximately normal when n is sufficiently large, which it is here, because $np = 196(0.092) = 18.032$ and $nq = 196(0.908) = 177.968$, and both are greater than 5.

$$\hat{p} = \frac{r}{n} = \frac{29}{196} = 0.1480$$

$$z = \frac{\hat{p} - p}{\sqrt{\dfrac{pq}{n}}} = \frac{0.1480 - 0.092}{\sqrt{\dfrac{0.092(0.908)}{196}}} = 2.71$$

(c) P value $= P(z > 2.71) = 0.0034$

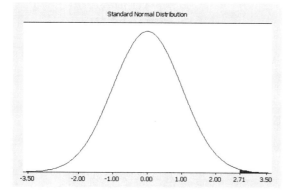

(d) Since a P value of $0.0034 \leq 0.05$, we reject H_0. Yes, the data are statistically significant.

(e) At the 5% level of significance, the data indicate that the population proportion of students with hypertension during final exams week is higher than 9.2%.

21. (a) $\alpha = 0.01$

$H_0 : p = 0.82$

$H_1 : p \neq 0.82$

(b) Use the standard normal distribution. The $\hat{p}$ distribution is approximately normal when n is sufficiently large, which it is here, because $np = 73(0.82) = 59.86$ and $nq = 73(0.18) = 13.14$, and both are greater than 5.

$$\hat{p} = \frac{r}{n} = \frac{56}{73} = 0.7671$$

$$z = \frac{\hat{p} - p}{\sqrt{\dfrac{pq}{n}}} = \frac{0.7671 - 0.82}{\sqrt{\dfrac{0.82(0.18)}{73}}} = -1.18$$

(c) P value $= 2P(z \leq -1.18) = 2(0.1190) = 0.2380$

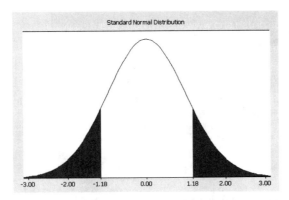

Standard Normal Distribution

(d) Since a P value of $0.2380 > 0.01$, we fail to reject H_0. No, the data are not statistically significant.

(e) At the 1% level of significance, the evidence is insufficient to indicate that the population proportion of extroverts among college student government leaders is different from 82%.

23. For a left-tailed test and $\alpha = 0.01$, critical value is $z_0 = -2.33$. The critical region consists of values less than -2.33. The sample test statistic $z = -2.65$ is in the critical region, so we reject H_0. Result is consistent with the P-value conclusion.

Section 8.4

1. Paired data are dependent.

3. $H_0 : \mu_d = 0$. We test that the mean difference is 0.

5. Here, $d.f. = n - 1$.

7. (a) Yes. The sample size is sufficiently large. Use a Student's t with $df = n - 1 = 35$.

(b) $H_0 : \mu_d = 0; \quad H_1 : \mu_d \neq 0$

(c) $t = \dfrac{\bar{d} - \mu_d}{s_d / \sqrt{n}} = \dfrac{0.8 - 0}{2 / \sqrt{36}} = 2.4$

(d) Table 6 gives $0.02 < P-\text{value} < 0.05$; TI-84 gives $P-\text{value} = 0.0218$

(e) Reject the null hypothesis as $P-\text{value} = 0.0218 < \alpha = 0.05$.

(f) At the 5% level of significance, the sample mean of differences is sufficiently different from 0 that we conclude the population mean of the differences is not zero.

9. **(a)** $\alpha = 0.05$

$H_0: \mu_d = 0$

$H_1: \mu_d \neq 0$

Since $\neq$ is in H_1, a two-tailed test is used.

(b) Use the Student's t distribution. Assume that d has a normal distribution or has a mound-shaped, symmetric distribution.

Pair	1	2	3	4	5	6	7	8
$d = B - A$	3	−2	5	4	10	−15	6	7

$\bar{d} = 2.25$, $s_d = 7.78$

$t = \dfrac{\bar{d} - \mu_d}{\frac{s_d}{\sqrt{n}}} = \dfrac{2.25 - 0}{\frac{7.78}{\sqrt{8}}} = 0.818$

(c) $d.f. = n - 1 = 8 - 1 = 7$

From Table 6 in Appendix II, 0.818 falls between entries 0.711 and 1.254. Using two-tailed areas, find that $0.250 < P$ value < 0.500. Using a TI-84, P value ≈ 0.4402.

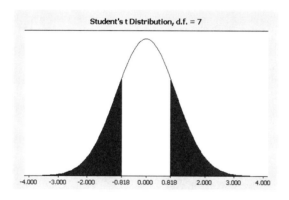

Student's t Distribution, d.f. = 7

(d) Since the P-value interval is > 0.05, we fail to reject H_0. No, the data are not statistically significant.

(e) At the 5% level of significance, the evidence is insufficient to claim a difference in population mean percentage increases for corporate revenue and CEO salary.

11. **(a)** $\alpha = 0.01$

$H_0: \mu_d = 0$

$H_1: \mu_d > 0$

Since $>$ is in H_1, a right-tailed test is used.

(b) Use the Student's t distribution. Assume that d has a normal distribution or has a mound-shaped, symmetric distribution.

Pair	1	2	3	4	5
$d = \text{Jan} - \text{April}$	35	9	26	−24	17

$$\bar{d} = 12.6, \ s_d = 22.66$$

$$t = \frac{\bar{d} - \mu_d}{\frac{s_d}{\sqrt{n}}} = \frac{12.6 - 0}{\frac{22.66}{\sqrt{5}}} = 1.243$$

(c) $d.f. = n - 1 = 5 - 1 = 4$

From Table 6 in Appendix II, 1.243 falls between entries 0.741 and 1.344. Using one-tailed areas, find that $0.125 < P$ value < 0.250. Using a TI-84, the P value ≈ 0.1408.

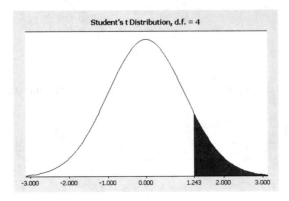

(d) Since the P-value interval is > 0.01, we fail to reject H_0. No, the data are not statistically significant.

(e) At the 1% level of significance, the evidence is insufficient to claim average peak wind gusts are higher in January.

13. (a) $\alpha = 0.05$

$H_0: \mu_d = 0$

$H_1: \mu_d > 0$

Since $>$ is in H_1, a right-tailed test is used.

(b) Use the Student's t distribution. Assume that d has a normal distribution or has a mound-shaped, symmetric distribution.

Pair	1	2	3	4	5	6	7	8
d = winter − summer	19	−4	17	7	9	0	−9	10

$$\bar{d} = 6.125, \ s_d = 9.83$$

$$t = \frac{\bar{d} - \mu_d}{\frac{s_d}{\sqrt{n}}} = \frac{6.125 - 0}{\frac{9.83}{\sqrt{8}}} = 1.76$$

(c) $d.f. = n - 1 = 8 - 1 = 7$

From Table 6 in Appendix II, 1.76 falls to the between entries 1.617 and 1.895. Using one-tailed areas, find that $0.05 < P$ value < 0.075. Using a TI-84, P value ≈ 0.0607.

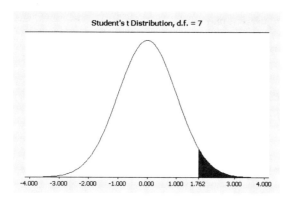

Student's t Distribution, d.f. = 7

(d) Since the *P*-value interval is > 0.05, we fail to reject H_0. No, the data are not statistically significant.

(e) At the 5% level of significance, the evidence is insufficient to indicate that the population average percentage of male wolves in winter is higher.

15. (a) $\alpha = 0.05$

$H_0: \mu_d = 0$

$H_1: \mu_d > 0$

Since > is in H_1, a right-tailed test is used.

(b) Use the Student's *t* distribution. Assume that *d* has a normal distribution or has a mound-shaped, symmetric distribution.

Pair	1	2	3	4	5	6	7	8
d = houses − hogans	5	2	22	−23	−4	−19	33	32

$\bar{d} = 6,\ s_d = 21.5$

$$t = \frac{\bar{d} - \mu_d}{\frac{s_d}{\sqrt{n}}} = \frac{6 - 0}{\frac{21.5}{\sqrt{8}}} = 0.789$$

(c) $d.f. = n - 1 = 8 - 1 = 7$

From Table 6 in Appendix II, 0.789 falls between entries 0.711 and 1.254. Use one-tailed areas to find that $0.125 < P$ value < 0.250. Using a TI-84, *P* value ≈ 0.2282.

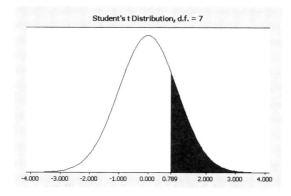

Student's t Distribution, d.f. = 7

(d) Since the *P*-value interval is > 0.05, we fail to reject H_0. No, the data are not statistically significant.

(e) At the 5% level of significance, the evidence is insufficient to show that the mean number of inhabited houses is greater than that of hogans.

17. (i) Use a calculator to verify. Rounded values are used in part (ii).

    (ii) (a) $\alpha = 0.05$

$$H_0 : \mu = 0$$

$$H_1 : \mu > 0$$

Since $>$ is in H_1, a right-tailed test is used.

(b) Use the Student's t distribution. The sample size is greater than 30.

$$\bar{d} = 2.472,\ s_d = 12.124$$

$$t = \frac{\bar{d} - \mu_d}{\frac{s_d}{\sqrt{n}}} = \frac{2.472 - 0}{\frac{12.124}{\sqrt{36}}} \approx 1.223$$

(c) $d.f. = n - 1 = 36 - 1 = 35$

From Table 6 in Appendix II, 1.223 falls between entries 1.170 and 1.306. Use one-tailed areas to find that $0.100 < P$ value < 0.125. Using a TI-84, P value ≈ 0.1147.

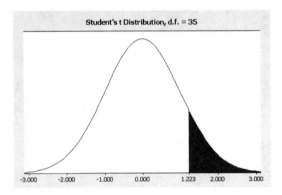

Student's t Distribution, d.f. = 35

(d) Since the P-value interval > 0.05, we fail to reject H_0. No, the data are not statistically significant.

(e) At the 5% level of significance, the evidence is insufficient to claim that the population mean cost of living index for housing is higher than that for groceries.

19. (a) $\alpha = 0.05$

$$H_0 : \mu_d = 0$$

$$H_1 : \mu_d > 0$$

Since $>$ is in H_1, a right-tailed test is used.

(b) Use the Student's t distribution. Assume that d has a normal distribution or has a mound-shaped, symmetric distribution.

Pair	1	2	3	4	5	6	7	8	9
$d = B - A$	7	−2	9	0	6	1	−3	−3	3

$$\bar{d} = 2.0,\ s_d = 4.5$$

$$t = \frac{\bar{d} - \mu_d}{\frac{s_d}{\sqrt{n}}} = \frac{2.0 - 0}{\frac{4.5}{\sqrt{9}}} = 1.33$$

(c) $d.f. = n - 1 = 9 - 1 = 8$

From Table 6 in Appendix II, 1.33 falls between entries 1.240 and 1.397. Use one-tailed areas to find that $0.100 < P$ value < 0.125. Using a TI-84, P value ≈ 0.1096.

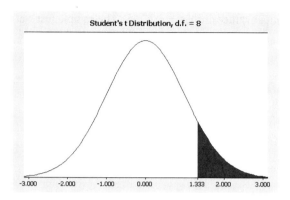

(d) Since the *P*-value interval is > 0.05, we fail to reject H_0. No, the data are not statistically significant.

(e) At the 5% level of significance, the evidence is insufficient to claim that the population score on the last round is higher than that on the first.

21. (a) $\alpha = 0.05$

$H_0: \mu_d = 0$

$H_1: \mu_d > 0$

Since > is in H_1, a right-tailed test is used.

(b) Use the Student's *t* distribution. Assume that *d* has a normal distribution or has a mound-shaped, symmetric distribution.

Pair	1	2	3	4	5	6	7	8
d = time 1 − time 5	1.4	1.7	−0.8	1.5	−0.5	−0.1	1.7	1.3

$\bar{d} = 0.775, \ s_d = 1.0539$

$t = \dfrac{\bar{d} - \mu_d}{\dfrac{s_d}{\sqrt{n}}} = \dfrac{0.775 - 0}{\dfrac{1.0539}{\sqrt{8}}} = 2.080$

(c) $d.f. = n - 1 = 8 - 1 = 7$

From Table 6 in Appendix II, 2.080 falls between entries 1.895 and 2.365. Use one-tailed areas to find $0.025 < P \text{ value} < 0.050$. Using a TI-84, $P \text{ value} \approx 0.0380$.

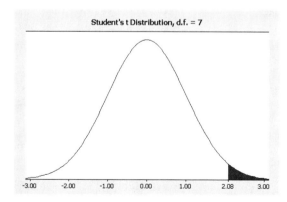

(d) Since the *P*-value interval is ≤ 0.05, we reject H_0. Yes, the data are statistically significant.

(e) At the 5% level of significance, the evidence is sufficient to claim that the population mean time for rats receiving larger rewards to climb the ladder is less.

23. For a two-tailed test with $\alpha = 0.05$ and $d.f. = 7$, critical values are $\pm t_0 = \pm 2.365$. The sample test statistic $t = 0.818$ is between -2.365 and 2.365, so we do not reject H_0. This conclusion is the same as that reached by the P-value method.

Section 8.5

1. **(a)** H_0 says the population means are equal.

   (b) $z = \dfrac{\bar{x}_1 - \bar{x}_2}{\sqrt{\dfrac{\sigma_1}{n_1} + \dfrac{\sigma_2}{n_2}}}$

   (c) $t = \dfrac{\bar{x}_1 - \bar{x}_2}{\sqrt{\dfrac{s_1^2}{n_1} + \dfrac{s_2^2}{n_2}}}$ with $df =$ smaller sample size $- 1$ or use Satterthwaite's formula

3. $H_0: \mu_1 = \mu_2$ or $H_0: \mu_1 - \mu_2 = 0$.

5. The best estimate is $\bar{p} = \dfrac{r_1 + r_2}{n_1 + n_2}$.

7. $H_1: \mu_1 > \mu_2$ or $H_1: \mu_1 - \mu_2 > 0$.

9. **(a)** A Student's t with $df = 48$. Samples are independent, the population standard deviations are not known, and the sample sizes are sufficiently large.

   (b) $H_0: \mu_1 = \mu_2$; $H_1: \mu_1 \neq \mu_2$

   (c) $\bar{x}_1 - \bar{x}_2 = 10 - 12 = -2$; $t = \dfrac{\bar{x}_1 - \bar{x}_2}{\sqrt{\dfrac{s_1^2}{n_1} + \dfrac{s_2^2}{n_2}}} = \dfrac{10 - 12}{\sqrt{\dfrac{3^2}{49} + \dfrac{4^2}{64}}} \approx -3.037$

   (d) Table 6 gives $0.001 < P - \text{value} < 0.01$; TI-84 gives $P - \text{value} = 0.0030$

   (e) Reject the null hypothesis as $P - \text{value} = 0.0030 < \alpha = 0.01$.

   (f) At the 1% level of significance, the sample evidence is sufficiently strong to reject the null hypothesis and conclude the population means are different.

11. **(a)** The standard normal distribution. Samples are independent, the population standard deviations are known, and the sample sizes are sufficiently large.

   (b) $H_0: \mu_1 = \mu_2$; $H_1: \mu_1 \neq \mu_2$

   (c) $\bar{x}_1 - \bar{x}_2 = 10 - 12 = -2$; $z = \dfrac{\bar{x}_1 - \bar{x}_2}{\sqrt{\dfrac{\sigma_1^2}{n_1} + \dfrac{\sigma_2^2}{n_2}}} = \dfrac{10 - 12}{\sqrt{\dfrac{3^2}{49} + \dfrac{4^2}{64}}} \approx -3.04$

   (d) Table 5 gives $P - \text{value} = 0.0024$

   (e) Reject the null hypothesis as $P - \text{value} = 0.0024 < \alpha = 0.01$.

   (f) At the 1% level of significance, the sample evidence is sufficiently strong to reject the null hypothesis and conclude the population means are different.

13. **(a)** $\bar{p} = \dfrac{45+70}{75+100} \approx 0.657$

(b) The standard normal distribution.

$n_1 \bar{p} = 75(0.657) = 49.275 > 5$

$n_2 \bar{p} = 100(0.657) = 65.7 > 5$

$n_1 \bar{q} = 75(0.343) = 25.725 > 5$

$n_2 \bar{q} = 100(0.343) = 34.3 > 5$

(c) $H_0 : p_1 = p_2 ; \quad H_1 : p_1 \neq p_2$

(d) $\hat{p}_1 - \hat{p}_2 = \dfrac{45}{75} - \dfrac{70}{100} = -0.1 ; \quad z = \dfrac{\hat{p}_1 - \hat{p}_2}{\sqrt{\dfrac{\overline{pq}}{n_1} + \dfrac{\overline{pq}}{n_2}}} = \dfrac{0.6 - 0.7}{\sqrt{\dfrac{0.657(0.343)}{75} + \dfrac{0.657(0.343)}{100}}} \approx -1.38$

(e) Table 5 gives $P - \text{value} = 0.1676$

(f) Do not reject the null hypothesis as $P - \text{value} = 0.1676 > \alpha = 0.05$.

(g) At the 5% level of significance, the difference between the sample proportions is too small to justify rejecting the null hypothesis that the probabilities are equal.

15. **(a)** $\alpha = 0.01$

$H_0 : \mu_1 = \mu_2$

$H_1 : \mu_1 > \mu_2$

(b) Use the standard normal distribution. We assume that both population distributions are approximately normal and that σ_1 and σ_2 are known.

$\bar{x}_1 - \bar{x}_2 = 2.8 - 2.1 = 0.7$

$z = \dfrac{(\bar{x}_1 - \bar{x}_2) - (\mu_1 - \mu_2)}{\sqrt{\dfrac{\sigma_1^2}{n_1} + \dfrac{\sigma_2^2}{n_2}}} = \dfrac{0.7 - 0}{\sqrt{\dfrac{0.5^2}{10} + \dfrac{0.7^2}{10}}} \approx 2.57$

(c) $P \text{ value} = P(z > 2.57) = 0.0051$

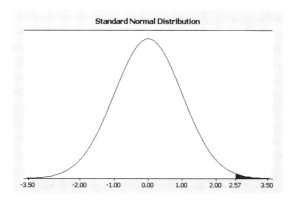

(d) Since $P \text{ value} = 0.0051 \leq 0.01$, we reject H_0. Yes, the data are statistically significant.

(e) At the 1% level of significance, the evidence is sufficient to indicate that the population mean REM sleep time for children is more than for adults.

17. (a) $\alpha = 0.05$

$$H_0: \mu_1 = \mu_2$$
$$H_1: \mu_1 \neq \mu_2$$

(b) Use the standard normal distribution. We assume that both population distributions are approximately normal and that σ_1 and σ_2 are known.

$$\bar{x}_1 - \bar{x}_2 = 4.9 - 4.3 = 0.6$$

$$z = \frac{(\bar{x}_1 - \bar{x}_2) - (\mu_1 - \mu_2)}{\sqrt{\frac{\sigma_1^2}{n_1} + \frac{\sigma_2^2}{n_2}}} = \frac{0.6 - 0}{\sqrt{\frac{1.5^2}{46} + \frac{1.2^2}{51}}} \approx 2.16$$

(c) P value $= 2P(z > 2.16) = 2(0.0154) = 0.0308$

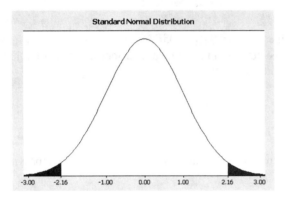

(d) Since P value $= 0.0308 \leq 0.05$, we reject H_0. Yes, the data are statistically significant.

(e) At the 5% level of significance, the evidence is sufficient to show that there is a difference between mean response regarding preference for camping or fishing.

19. (i) Use a calculator to verify. Use rounded results to compute t.

(ii) (a) $\alpha = 0.01$

$$H_0: \mu_1 = \mu_2$$
$$H_1: \mu_1 < \mu_2$$

(b) Use the Student's t distribution. We assume that both population distributions are approximately normal and that σ_1 and σ_2 are unknown.

$$\bar{x}_1 - \bar{x}_2 = 3.51 - 3.87 = -0.36$$

$$t = \frac{(\bar{x}_1 - \bar{x}_2) - (\mu_1 - \mu_2)}{\sqrt{\frac{s_1^2}{n_1} + \frac{s_2^2}{n_2}}} = \frac{-0.36 - 0}{\sqrt{\frac{0.81^2}{10} + \frac{0.94^2}{12}}} \approx -0.965$$

(c) Since $n_1 < n_2$, $d.f. = n_1 - 1 = 10 - 1 = 9$

In Table 6 in Appendix II, 0.965 falls between entries 0.703 and 1.230. Use one-tailed areas to find that $0.125 < P$ value < 0.250. Using a TI-84, $d.f. \approx 19.96$, and P value ≈ 0.1731.

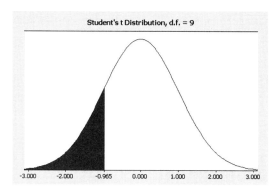

(d) Since the *P*-value interval > 0.01, we fail to reject H_0. No, the data are not statistically significant.

(e) At the 1% level of significance, the evidence is insufficient to indicate that violent crime in the Rocky Mountain region is higher than in New England.

21. (a) $\alpha = 0.05$

$H_0: \mu_1 = \mu_2$

$H_1: \mu_1 \neq \mu_2$

(b) Use the Student's *t* distribution. Both sample sizes are large, $n_i \geq 30$, and σ_1 and σ_2 are unknown.

$\bar{x}_1 - \bar{x}_2 = 344.5 - 354.2 = -9.7$

$$t = \frac{(\bar{x}_1 - \bar{x}_2) - (\mu_1 - \mu_2)}{\sqrt{\dfrac{s_1^2}{n_1} + \dfrac{s_2^2}{n_2}}} = \frac{-9.7 - 0}{\sqrt{\dfrac{49.1^2}{30} + \dfrac{50.9^2}{30}}} \approx -0.751$$

(c) Since $n_1 = n_2$, $d.f. = n - 1 = 30 - 1 = 29$.

From Table 6 in Appendix II, 0.751 falls between entries 0.683 and 1.174. Use two-tailed areas to find $0.250 < P$ value < 0.500. Using a TI-84, $d.f. \approx 57.92$, and *P* value ≈ 0.4556.

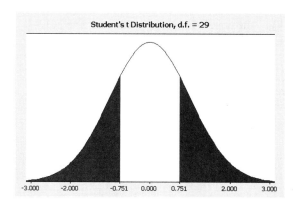

(d) Since the *P*-value interval > 0.05, we fail to reject H_0. No, the data are not statistically significant.

(e) At the 5% level of significance, the evidence is insufficient to indicate there is a difference between the control and experimental groups in the mean score on the vocabulary portion of the test.

23. (i) Use a calculator to verify. Use rounded results to compute *t*.

(ii) (a) $\alpha = 0.05$

$H_0: \mu_1 = \mu_2$

$H_1: \mu_1 \neq \mu_2$

(b) Use the Student's t distribution. We assume that the population distributions for both are approximately normal and mound-shaped and that σ_1 and σ_2 are unknown.

$$\bar{x}_1 - \bar{x}_2 = 4.75 - 3.93 = 0.82$$

$$t = \frac{(\bar{x}_1 - \bar{x}_2) - (\mu_1 - \mu_2)}{\sqrt{\dfrac{s_1^2}{n_1} + \dfrac{s_2^2}{n_2}}} = \frac{0.82 - 0}{\sqrt{\dfrac{2.82^2}{16} + \dfrac{2.43^2}{15}}} \approx 0.869$$

(c) Since $n_2 < n_1$, $d.f. = n_2 - 1 = 15 - 1 = 14$.

From Table 6 in Appendix II, 0.869 falls between entries 0.692 and 1.200. Use two-tailed areas to find $0.250 < P\text{ value} < 0.500$. Using a TI-84, $d.f. \approx 28.81$, and $P\text{ value} \approx 0.3921$.

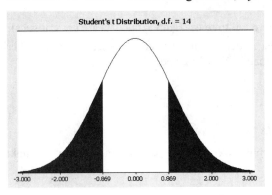

(d) Since the P-value interval is > 0.05, we fail to reject H_0. No, the data are not statistically significant.

(e) At the 5% level of significance, the evidence is insufficient to indicate that there is a difference in the mean number of cases of fox rabies between the two regions.

25. (i) Use a calculator to verify. Use rounded results to compute t.

(ii) (a) $\alpha = 0.05$

$H_0: \mu_1 = \mu_2$

$H_1: \mu_1 \neq \mu_2$

(b) Use the Student's t distribution. We assume the population distributions for both are approximately normal and mound-shaped and that σ_1 and σ_2 are unknown.

$$\bar{x}_1 - \bar{x}_2 = 4.86 - 6.5 = -1.64$$

$$t = \frac{(\bar{x}_1 - \bar{x}_2) - (\mu_1 - \mu_2)}{\sqrt{\dfrac{s_1^2}{n_1} + \dfrac{s_2^2}{n_2}}} = \frac{-1.64 - 0}{\sqrt{\dfrac{3.18^2}{7} + \dfrac{2.88^2}{8}}} \approx -1.041$$

(c) Since $n_1 < n_2$, $d.f. = n_1 - 1 = 7 - 1 = 6$.

From Table 6 in Appendix II, 1.041 falls between entries 0.718 and 1.273. Use two-tailed areas to find $0.250 < P\text{ value} < 0.500$. Using a TI-84, $d.f. \approx 12.28$, and $P\text{ value} \approx 0.3179$.

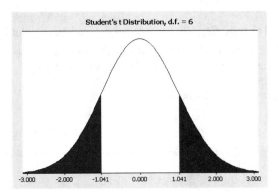

(d) Since the P-value interval is > 0.05, we fail to reject H_0. No, the data are not statistically significant.

(e) At the 5% level of significance, the evidence is insufficient to indicate that the mean time lost owing to hot tempers is different from that lost owing to technical workers' attitudes.

27. (a) $d.f. \approx \dfrac{\left(\dfrac{s_1^2}{n_1} + \dfrac{s_2^2}{n_2}\right)^2}{\dfrac{1}{n_1-1}\left(\dfrac{s_1^2}{n_1}\right)^2 + \dfrac{1}{n_2-1}\left(\dfrac{s_2^2}{n_2}\right)^2} = \dfrac{\left(\dfrac{0.81^2}{10} + \dfrac{0.94^2}{12}\right)^2}{\dfrac{1}{9}\left(\dfrac{0.81^2}{10}\right)^2 + \dfrac{1}{11}\left(\dfrac{0.94^2}{12}\right)^2} \approx 19.96$

(*Note:* Some software will truncate this to 19.)

(b) In Problem 19, $d.f. = n_1 - 1 = 10 - 1 = 9$.

The convention of using the smaller of $n_1 - 1$ and $n_2 - 1$ leads to a $d.f.$ that is always less than or equal to that computed by Satterthwaite's formula.

29. (a) $\alpha = 0.05$

$H_0 : p_1 = p_2$

$H_1 : p_1 \neq p_2$

(b) Use the standard normal distribution. The number of trials is sufficiently large because $n_1 \bar{p}$, $n_1 \bar{q}$, $n_2 \bar{p}$, and $n_2 \bar{q}$ are each larger than 5.

$\bar{p} = \dfrac{r_1 + r_2}{n_1 + n_2} = \dfrac{59 + 56}{220 + 175} \approx 0.2911$

$\bar{q} = 1 - \bar{p} = 1 - 0.2911 = 0.7089$

$\hat{p}_1 = \dfrac{r_1}{n_1} = \dfrac{59}{220} \approx 0.268, \ \hat{p}_2 = \dfrac{r_2}{n_2} = \dfrac{56}{175} = 0.32$

$\hat{p}_1 - \hat{p}_2 = 0.268 - 0.32 = -0.052$

$z = \dfrac{\hat{p}_1 - \hat{p}_2}{\sqrt{\dfrac{\overline{pq}}{n_1} + \dfrac{\overline{pq}}{n_2}}} = \dfrac{-0.052}{\sqrt{\dfrac{0.2911(0.7089)}{220} + \dfrac{0.2911(0.7089)}{175}}} \approx -1.13$

(c) $P \text{ value} = 2P(z < -1.13) = 2(0.1292) = 0.2584$

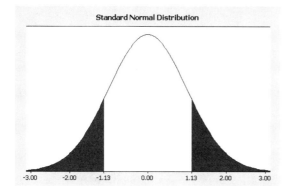

(d) Since the P value $= 0.2584 > 0.05$, we fail to reject H_0. No, the data are not statistically significant.

(e) At the 1% level of significance, there is insufficient evidence to conclude that the population proportion of women favoring more tax dollars for the arts is different from the proportion of men.

31. **(a)** $\alpha = 0.01$

$H_0\colon p_1 = p_2$

$H_1\colon p_1 \neq p_2$

(b) Use the standard normal distribution. The number of trials is sufficiently large because $n_1\overline{p}$, $n_1\overline{q}$, $n_2\overline{p}$, and $n_2\overline{q}$ are each larger than 5.

$$\overline{p} = \frac{r_1 + r_2}{n_1 + n_2} = \frac{12 + 7}{153 + 128} = 0.0676$$

$$\overline{q} = 1 - \overline{p} = 1 - 0.0676 = 0.9324$$

$$\hat{p}_1 = \frac{r_1}{n_1} = \frac{12}{153} \approx 0.0784, \ \hat{p}_2 = \frac{r_2}{n_2} = \frac{7}{128} \approx 0.0547$$

$$\hat{p}_1 - \hat{p}_2 = 0.0784 - 0.0547 = 0.0237$$

$$z = \frac{\hat{p}_1 - \hat{p}_2}{\sqrt{\dfrac{\overline{pq}}{n_1} + \dfrac{\overline{pq}}{n_2}}} = \frac{0.0237}{\sqrt{\dfrac{0.0676(0.9324)}{153} + \dfrac{0.0676(0.9324)}{128}}} \approx 0.79$$

(c) P value $= 2P(z > 0.79) = 2(0.2148) = 0.4296$

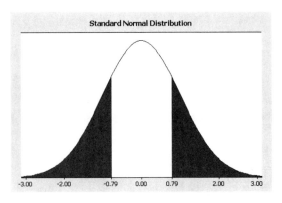

Standard Normal Distribution

(d) Since the P value $= 0.4296 > 0.01$, we fail to reject H_0. No, the data are not statistically significant.

(e) At the 1% level of significance, there is insufficient evidence to conclude that the population proportion of high-school dropouts on Oahu is different from that of Sweetwater County.

33. **(a)** $\alpha = 0.01$

$H_0\colon p_1 = p_2$

$H_1\colon p_1 < p_2$

(b) Use the standard normal distribution. The number of trials is sufficiently large because $n_1\overline{p}$, $n_1\overline{q}$, $n_2\overline{p}$, and $n_2\overline{q}$ are each larger than 5.

$$\overline{p} = \frac{r_1 + r_2}{n_1 + n_2} = \frac{37 + 47}{100 + 100} = 0.42$$

$$\overline{q} = 1 - \overline{p} = 1 - 0.42 = 0.58$$

$$\hat{p}_1 = \frac{r_1}{n_1} = \frac{37}{100} = 0.37, \ \hat{p}_2 = \frac{r_2}{n_2} = \frac{47}{100} = 0.47$$

$$\hat{p}_1 - \hat{p}_2 = 0.37 - 0.47 = -0.10$$

$$z = \frac{\hat{p}_1 - \hat{p}_2}{\sqrt{\dfrac{\overline{pq}}{n_1} + \dfrac{\overline{pq}}{n_2}}} = \frac{-0.10}{\sqrt{\dfrac{0.42(0.58)}{100} + \dfrac{0.42(0.58)}{100}}} \approx -1.43$$

(c) P value $= P(z < -1.43) \approx 0.0764$

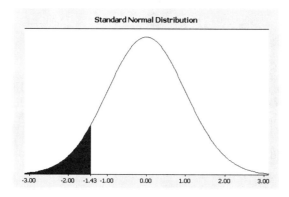

(d) Since the P value $= 0.0764 > 0.01$, we fail to reject H_0. No, the data are not statistically significant.

(e) At the 1% level of significance, there is insufficient evidence to conclude that the population proportion of adults believing in extraterrestrials who attended college is higher than the proportion who did not attend college.

35. (a) $\alpha = 0.05$

$H_0 : p_1 = p_2$

$H_1 : p_1 < p_2$

(b) Use the standard normal distribution. The number of trials is sufficiently large because $n_1\overline{p}$, $n_1\overline{q}$, $n_2\overline{p}$, and $n_2\overline{q}$ are each larger than 5.

$$\overline{p} = \frac{r_1 + r_2}{n_1 + n_2} = \frac{45 + 71}{250 + 280} \approx 0.2189$$

$$\overline{q} = 1 - \overline{p} = 1 - 0.2189 = 0.7811$$

$$\hat{p}_1 = \frac{r_1}{n_1} = \frac{45}{250} = 0.18, \ \hat{p}_2 = \frac{r_2}{n_2} = \frac{71}{280} \approx 0.2536$$

$$\hat{p}_1 - \hat{p}_2 = 0.18 - 0.2536 \approx -0.074$$

$$z = \frac{\hat{p}_1 - \hat{p}_2}{\sqrt{\dfrac{\overline{pq}}{n_1} + \dfrac{\overline{pq}}{n_2}}} = \frac{-0.074}{\sqrt{\dfrac{0.2189(0.7811)}{250} + \dfrac{0.2189(0.7811)}{280}}} \approx -2.04$$

(c) P value $= P(z < -2.04) = 0.0207$

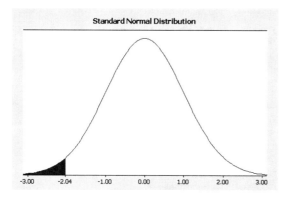

(d)   Since the P value $= 0.0207 < 0.05$, we reject H_0. Yes, the data are statistically significant.

(e)   At the 5% level of significance, there is sufficient evidence to conclude that the population proportion of trusting people in Chicago is higher in the older group.

37. $H_0: \mu_1 = \mu_2$; $H_1: \mu_1 < \mu_2$; for $d.f. = 9$, $\alpha = 0.01$ in the *one-tail area* row, the critical value $t_0 = -2.821$. Sample test statistic $t = -0.965$ is not in the critical region, fail to reject H_0. This result is consistent with the P-value method.

Chapter Review Problems

1.   Look at the original x distribution. If it is normal or $n \geq 30$, and σ is known, use the standard normal distribution. If the x distribution is mound-shaped or $n \geq 30$, and σ is unknown, use the Student's t distribution. The degrees of freedom are determined by the application.

3.   A larger sample size increases the value of $\left| z \right|$ or $\left| t \right|$.

5.   (a) $\alpha = 0.05$

   $H_0: \mu = 11.1$

   $H_1: \mu \neq 11.1$

   Since $\neq$ is in H_1, use a two-tailed test.

   (b)   Use the standard normal distribution. We assume that x has a normal distribution with known standard deviation σ. Note that μ is given in the null hypothesis.

   $$z = \frac{\overline{x} - \mu}{\frac{\sigma}{\sqrt{n}}} = \frac{10.8 - 11.1}{\frac{0.6}{\sqrt{36}}} = -3.00$$

   (*Note:* 600 miles = 0.6 thousand miles)

   (c) P value $= 2P(z < -3.00) = 2(0.0013) = 0.0026$

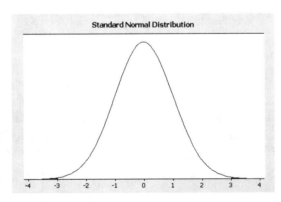

Standard Normal Distribution

   (d)   Since a P value of $0.0026 \leq 0.05$, we reject H_0. Yes, the data are statistically significant.

   (e)   At the 5% level of significance, the evidence is sufficient to say that the miles driven per vehicle in Chicago is different from the national average.

7. **(a)** $\alpha = 0.01$

    $H_0: \mu = 0.8$ A

    $H_1: \mu > 0.8$ A

    (b) Use the Student's t distribution with $d.f. = n - 1 = 9 - 1 = 8$. We assume that x has a distribution that is approximately normal and that σ is unknown.

    $$t = \frac{\bar{x} - \mu}{\frac{s}{\sqrt{n}}} = \frac{1.4 - 0.8}{\frac{0.41}{\sqrt{9}}} \approx 4.390$$

    (c) For $d.f. = 8$, 4.390 falls between entries 3.355 and 5.041. Using one-tailed areas, $0.0005 < P$ value < 0.005. Using a TI-84, P value ≈ 0.0012.

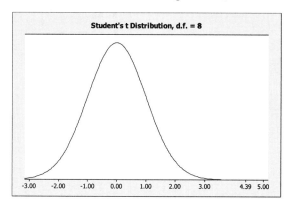

Student's t Distribution, d.f. = 8

    (d) Since the entire P-value interval ≤ 0.01, we reject H_0. Yes, the data are statistically significant.

    (e) At the 5% level of significance, the evidence is sufficient to say that the Toylot claim of 0.8 A is too low.

9. **(a)** $\alpha = 0.01$

    $H_0: p = 0.60$

    $H_1: p < 0.60$

    (b) Use the standard normal distribution. The $\hat{p}$ distribution is approximately normal when n is sufficiently large, which it is here, because $np = 90(0.6) = 54$ and $nq = 90(0.4) = 36$, and both are greater than 5.

    $$\hat{p} = \frac{r}{n} = \frac{40}{90} = 0.4444$$

    $$z = \frac{\hat{p} - p}{\sqrt{\frac{pq}{n}}} = \frac{0.4444 - 0.60}{\sqrt{\frac{0.60(0.40)}{90}}} = -3.01$$

    (c) P value $= P(z < -3.01) = 0.0013$

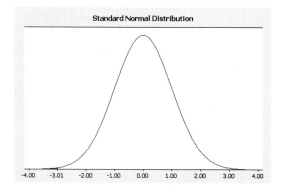

Standard Normal Distribution

(d) Since a P value of $0.0013 \leq 0.01$, we reject H_0. Yes, the data are statistically significant.

(e) At the 1% level of significance, the evidence is sufficient to show the mortality rate has dropped.

11. (a) $\alpha = 0.01$

$H_0: \mu = 40$

$H_1: \mu > 40$

Since $>$ is in H_1, use a right-tailed test.

(b) Use the standard normal distribution. We assume that x has a normal distribution with known standard deviation σ. Note that μ is given in the null hypothesis.

$$z = \frac{\bar{x} - \mu}{\frac{\sigma}{\sqrt{n}}} = \frac{43.1 - 40}{\frac{9}{\sqrt{94}}} \approx 3.34$$

(c) P value $= P(z > 3.34) = 0.0004$

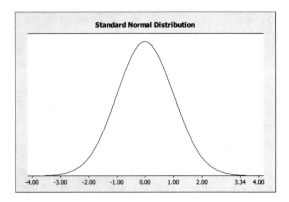

(d) Since a P value of $0.0004 \leq 0.01$, we reject H_0. Yes, the data are statistically significant.

(e) At the 1% level of significance, the evidence is sufficient to say that the population average number of matches is larger than 40.

13. (a) $\alpha = 0.05$

$H_0: \mu_1 = \mu_2$

$H_1: \mu_1 \neq \mu_2$

(b) Use the Student's t distribution. We assume that both population distributions are approximately normal and that σ_1 and σ_2 are unknown.

$\bar{x}_1 - \bar{x}_2 = 3.0 - 2.7 = 0.3$

$$t = \frac{(\bar{x}_1 - \bar{x}_2) - (\mu_1 - \mu_2)}{\sqrt{\frac{s_1^2}{n_1} + \frac{s_2^2}{n_2}}} = \frac{0.3 - 0}{\sqrt{\frac{0.8^2}{55} + \frac{0.9^2}{51}}} \approx 1.808$$

(c) Since $n_2 < n_1$, $d.f. = n_2 - 1 = 51 - 1 = 50$.

In Table 6 in Appendix II, 1.808 falls between entries 1.676 and 2.009. Use two-tailed areas to find that $0.050 < P$ value < 0.100. Using a TI-84, $d.f. \approx 100.27$, and P value ≈ 0.0735.

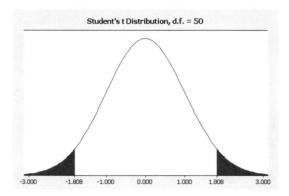

Student's t Distribution, d.f. = 50

-3.000   -1.808   -1.000   0.000   1.000   1.808   3.000

(d) Since the *P*-value interval > 0.05, we fail to reject H_0. No, the data are not statistically significant.

(e) At the 5% level of significance, the evidence is insufficient to show that there is a difference in population mean length of the two types of projectile points.

15. (a) $\alpha = 0.05$

$H_0: \mu = 7$ oz

$H_1: \mu \neq 7$ oz

(b) Use the Student's *t* distribution with *d.f.* = $n - 1 = 8 - 1 = 7$. We assume that *x* has a distribution that is approximately normal and that σ is unknown.

$$t = \frac{\bar{x} - \mu}{\dfrac{s}{\sqrt{n}}} = \frac{7.3 - 7}{\dfrac{0.5}{\sqrt{8}}} \approx 1.697$$

(c) For *d.f.* = 7, 1.697 falls between entries 1.617 and 1.895. Using two-tailed areas, $0.100 < P$ value < 0.150. Using a TI-84, *P* value ≈ 0.1335.

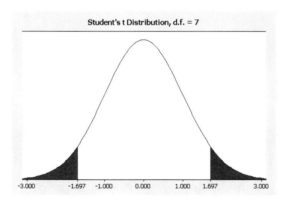

Student's t Distribution, d.f. = 7

-3.000   -1.697   -1.000   0.000   1.000   1.697   3.000

(d) Since the entire *P*-value interval > 0.05, we fail to reject H_0. No, the data are not statistically significant.

(e) At the 5% level of significance, the evidence is insufficient to show that the population mean amount of coffee per cup is different from 7 ounces.

17. (a) $\alpha = 0.05$

$H_0: \mu_d = 0$

$H_1: \mu_d < 0$

(b) Use the Student's *t* distribution. Assume that *d* has a normal distribution or has a mound-shaped, symmetric distribution.

Pair	1	2	3	4	5
d = before − after	−6.4	−7.2	−8.6	−3.8	1.3

$$\bar{d} = -4.94, \, s_d = 3.90$$

$$t = \frac{\bar{d} - \mu_d}{\frac{s_d}{\sqrt{n}}} = \frac{-4.94 - 0}{\frac{3.90}{\sqrt{5}}} = -2.832$$

(c) $d.f. = n - 1 = 5 - 1 = 4$

From Table 6 in Appendix II, 2.832 falls between entries 2.776 and 3.747. Use one-tailed areas to find that $0.010 < P$ value < 0.025. Using a TI-84, P value ≈ 0.0236.

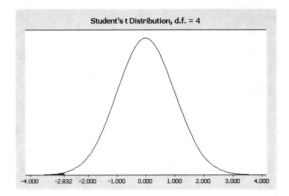

(d) Since the P-value interval is ≤ 0.05, we reject H_0. Yes, the data are statistically significant.

(e) At the 5% level of significance, there is insufficient evidence to claim that the injection system lasts less than an average of 48 months.

Chapter 9: Correlation and Regression

Section 9.1

1. The explanatory variable is graphed along the horizontal or x axis. The response variable is graphed along the vertical or y axis.

3. If two variables are negatively correlated, then the response variable will decrease as the explanatory variable increases.

5. (a) The points seem close to a straight line, so there is moderate linear correlation.
   (b) A straight line does not fit the data well, so there is no linear correlation.
   (c) The points seem very close to a straight line, so there is high linear correlation.

7. (a) No. The correlation coefficient is a mathematical tool for measuring the strength of a linear relationship between two variables. As such, it makes no implication about cause or effect. Just because two variables tend to increase or decrease together does not mean a change in one is *causing* a change in the other. A strong correlation between x and y is sometimes due to the presence of lurking variables.
   (b) Increasing population might be a lurking variable causing both variables to increase.

9. (a) No. The correlation coefficient is a mathematical tool for measuring the strength of a linear relationship between two variables. As such, it makes no implication about cause or effect. Just because two variables tend to increase or decrease together does not mean a change in one is *causing* a change in the other. A strong correlation between x and y is sometimes due to the presence of lurking variables.
   (b) One lurking variable for the increase in average annual income is inflation. Better training might be a lurking variable responsible for shorter times to run the mile.

11. The correlation coefficient is moderate and negative, suggesting that as gasoline prices increase, consumption decreases. It is risky to apply these results to gasoline prices higher than $5.30 per gallon as the study only included prices up to $5.30. We would not be sure if the relationship holds for prices beyond $5.30.

13. (a)

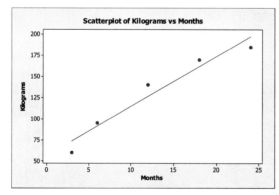

   (b) Since the points lie close to a straight line, the correlation is strong and positive.
   (c) $$r = \frac{n\sum xy - (\sum x)(\sum y)}{\sqrt{n\sum x^2 - (\sum x)^2}\sqrt{n\sum y^2 - (\sum y)^2}} = \frac{5(9,930) - (63)(650)}{\sqrt{5(1,089) - (63)^2}\sqrt{5(95,350) - (650)^2}} \approx 0.972$$

   Since r is positive, y should tend to increase as x increases.

15. (a)

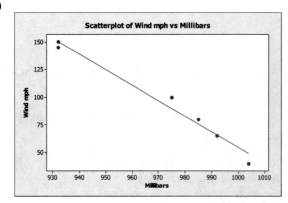

(b) Since the points lie very close to a straight line, the correlation is strong and negative.

(c) $r = \dfrac{n\sum xy - \left(\sum x\right)\left(\sum y\right)}{\sqrt{n\sum x^2 - \left(\sum x\right)^2}\sqrt{n\sum y^2 - \left(\sum y\right)^2}} = \dfrac{6(556{,}315) - (5{,}823)(580)}{\sqrt{6(5{,}655{,}779) - (5{,}823)^2}\sqrt{6(65{,}750) - (580)^2}} \approx -0.990$

Since r is negative, y should tend to decrease as x increases.

17. (a)

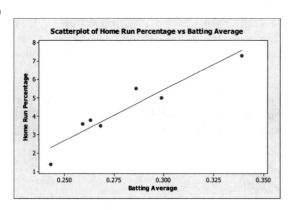

(b) Since the points are very close to a straight line, the correlation is strong and positive.

(c) $r = \dfrac{n\sum xy - \left(\sum x\right)\left(\sum y\right)}{\sqrt{n\sum x^2 - \left(\sum x\right)^2}\sqrt{n\sum y^2 - \left(\sum y\right)^2}} = \dfrac{7(8.753) - (1.957)(30.1)}{\sqrt{7(0.553) - (1.957)^2}\sqrt{7(150.15) - (30.1)^2}} \approx 0.948$

Since r is positive, y should tend to increase as x increases.

19.   (a)

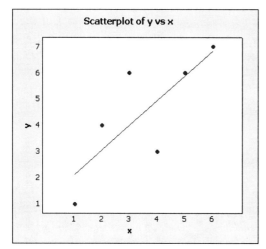

The axes are scaled equally.

(b)

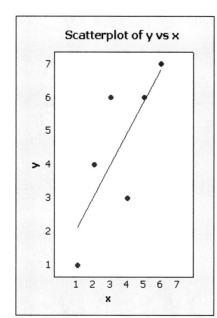

The *y* axis is twice the *x* axis.

(c)

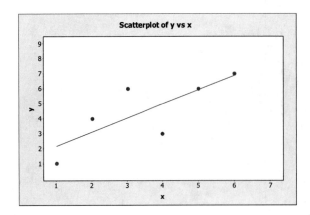

The x axis is twice the y axis.

(d) Stretching the scale on the y axis makes the line appear steeper. Shrinking the scale on the y axis makes the line appear flatter. The slope of the line does not change. Only the appearance of the slope changes as the scale of the y axis changes.

21. (a) $r \approx 0.972$ with $n = 5$

Since $|0.972| = 0.972 \geq 0.88$ for $\alpha = 0.05$, r is significant, and we conclude that age and weight of Shetland ponies are correlated.

(b) $r \approx -0.990$ with $n = 6$

Since $|-0.990| = 0.990 \geq 0.92$ for $\alpha = 0.01$, r is significant, and we conclude that the lowest barometric pressure reading and maximum wind speed for cyclones are correlated.

23. (a)

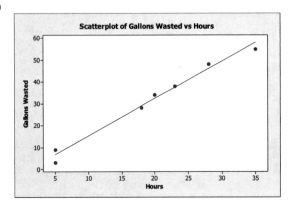

$$ r = \frac{n\sum xy - \left(\sum x\right)\left(\sum y\right)}{\sqrt{n\sum x^2 - \left(\sum x\right)^2}\,\sqrt{n\sum y^2 - \left(\sum y\right)^2}} = \frac{8(6{,}067) - (154)(249)}{\sqrt{8(3{,}712) - (154)^2}\,\sqrt{8(9{,}959) - (249)^2}} \approx 0.991 $$

(b) For variables based on averages, $\bar{x} = 19.25$ h; $s_x \approx 10.33$ h; $\bar{y} = 31.13$ gal; $s_y \approx 17.76$ gal.

For variables based on single individuals, $\bar{x} = 20.13$ h; $s_x \approx 13.84$ h; $\bar{y} = 31.87$ gal; $s_y \approx 25.18$.

Dividing by larger numbers results in a smaller value.

(c)

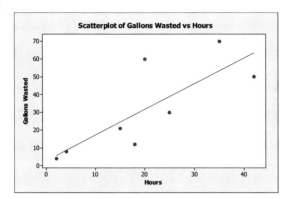

$$r = \frac{8(7,071) - (161)(255)}{\sqrt{8(4,583) - (161)^2}\sqrt{8(12,565) - (255)^2}} \approx 0.794$$

(d) $0.991 > 0.791$

Yes, by the central limit theorem, the $\bar{x}$ distribution has a smaller standard deviation than the corresponding x distribution.

Section 9.2

Note: **In Sections 9.2, 9.3, and 9.4, answers may vary slightly depending on rounding considerations.**

1. The slope is $b = -2$. When x increases by 1 unit, y decreases by 2 units.

3. We are extrapolating. Extrapolation is dangerous because the pattern of data may change outside the x range.

5. **(a)** $\hat{y} \approx 318.16 - 30.878x$
(b) There will be about 31 few frost-free days (30.878).
(c) Here, $r \approx -0.981$. Note that if the slope is negative, then r is also negative.
(d) 96.3% of the variation can be explained, which leaves 3.7% of the variation unexplained.

7. **(a)**

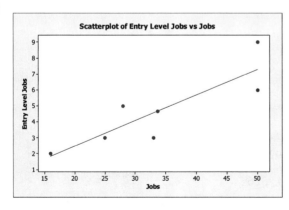

(b) Use a calculator to verify.

(c) $\bar{x} = \dfrac{\sum x}{n} = \dfrac{202}{6} \approx 33.67$ jobs

$\bar{y} = \dfrac{\sum y}{n} = \dfrac{28}{6} \approx 4.67$ entry-level jobs

$b = \dfrac{n\sum xy - (\sum x)(\sum y)}{n\sum x^2 - (\sum x)^2} = \dfrac{6(1,096) - (202)(28)}{6(7,754) - (202)^2} \approx 0.161$

$a = \bar{y} - b\bar{x} \approx 4.67 - (0.161)(33.67) \approx -0.748$

$\hat{y} = a + bx$ or $\hat{y} \approx -0.748 + 0.161x$

(d)   See figure of part (a).

(e) $r^2 = (0.860)^2 \approx 0.740$

This means that 74.0% of the variation in y = number of entry-level jobs can be explained by the corresponding variation in x = total number of jobs using the least-squares line.
100% − 74.0% = 26.0% of the variation is unexplained.

(f)   Use $x = 40$.

$\hat{y} = -0.748 + 0.161(40)$

$\hat{y} = 5.69$ entry-level jobs

9.   **(a)**

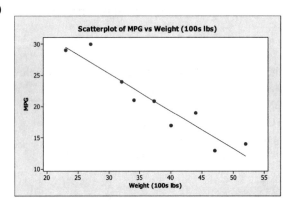

(b)   Use a calculator to verify.

(c) $\bar{x} = \dfrac{\sum x}{n} = \dfrac{299}{8} = 37.375$

$\bar{y} = \dfrac{\sum y}{n} = \dfrac{167}{8} = 20.875$

$b = \dfrac{n\sum xy - (\sum x)(\sum y)}{n\sum x^2 - (\sum x)^2} = \dfrac{8(5,814) - (299)(167)}{8(11,887) - (167)^2} \approx -0.6007$

$a = \bar{y} - b\bar{x} = 20.875 - (-0.6007)(37.375) \approx 43.326$

$\hat{y} = a + bx$ or $\hat{y} = 43.326 - 0.6007x$

(d)   See figure of part (a).

(e) $r^2 = (-0.946)^2 \approx 0.895$

This means that 89.5% of the variation in y = miles per gallon can be explained by the corresponding variation in x = weight using the least-squares line. 100% − 89.5% = 10.5% of the variation is unexplained.

(f) Use $x = 38$.

$\hat{y} = 43.326 - 0.6007(38) = 20.5$ mpg

11. (a)

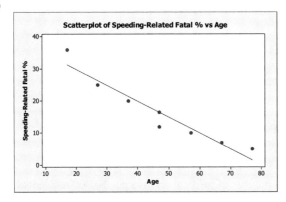

(b) Use a calculator to verify.

(c) $\bar{x} = \dfrac{\sum x}{n} = \dfrac{329}{7} = 47$ years

$\bar{y} = \dfrac{\sum y}{n} = \dfrac{115}{7} \approx 16.43\%$

$b = \dfrac{n\sum xy - \left(\sum x\right)\left(\sum y\right)}{n\sum x^2 - \left(\sum x\right)^2} =, \dfrac{7(4015) - (329)(115)}{7(18,263) - (329)^2} \approx -0.496$

$a = \bar{y} - b\bar{x} \approx 16.43 - (-0.496)(47) \approx 39.761$

$\hat{y} = a + bx$ or $\hat{y} = 39.761 - 0.496x$

(d) See the figure in part (a).

(e) $r^2 = (-0.959)^2 \approx 0.920$

This means that 92% of the variation in y = percentage of all fatal accidents due to speeding can be explained by the corresponding variation in x = age in years of a licensed automobile driver using the least-squares line. $100\% - 92\% = 8\%$ of the variation is unexplained.

(f) Use $x = 25$.

$\hat{y} = 39.761 - 0.496(25) \approx 27.36\%$

13. (a)

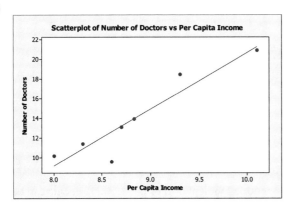

(b) Use a calculator to verify.

(c) $\bar{x} = \dfrac{\sum x}{n} = \dfrac{53}{6} \approx \$8.83$ thousand

$\bar{y} = \dfrac{\sum y}{n} = \dfrac{83.7}{6} = 13.95$ physicians per 10,000

$b = \dfrac{n\sum xy - (\sum x)(\sum y)}{n\sum x^2 - (\sum x)^2} = \dfrac{6(755.89) - (53)(83.7)}{6(471.04) - (53)^2} \approx 5.756$

$a = \bar{y} - b\bar{x} \approx 13.95 - 5.756(8.83) \approx -36.898$

$\hat{y} = a + bx$ or $\hat{y} \approx -36.898 + 5.756x$

(d) See the figure in part (a).

(e) $r^2 = (0.934)^2 \approx 0.872$

This means that 87.2% of the variation in y = number of medical doctors per 10,000 residents can be explained by the corresponding variation in x = per capita income in thousands of dollars using the least-squares line. 100% – 87.2% = 12.8% of the variation is unexplained.

(f) Use $x = 10$.

$\hat{y} = -36.898 + 5.756(10) \approx 20.7$ physicians per 10,000 residents

15. (a)

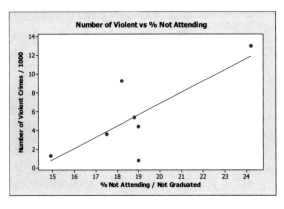

(b) Use a calculator to verify.

(c) $\bar{x} = \dfrac{\sum x}{n} = \dfrac{112.8}{6} = 18.8\%$

$\bar{y} = \dfrac{\sum y}{n} = \dfrac{32.4}{6} = 5.4$ crimes per 1,000

$b = \dfrac{n\sum xy - (\sum x)(\sum y)}{n\sum x^2 - (\sum x)^2} = \dfrac{6(665.03) - (112.8)(32.4)}{6(2,167.14) - (112.8)^2} \approx 1.202$

$a = \bar{y} - b\bar{x} \approx 5.4 - 1.202(18.8) \approx -17.204$

$\hat{y} = a + bx$ or $\hat{y} = -17.204 + 1.202x$

(d) See the figure in part (a).

(e) $r^2 = (0.764)^2 \approx 0.584$

This means that 58.4% of the variation in y = reported violent crimes per 1,000 residents can be explained by the corresponding variation in x = percentage of 16- to 19-year-olds not in school and not high-school graduates using the least-squares line. 100% – 58.4% = 41.6% of the variation is unexplained.

(f) Use $x = 24$.

$\hat{y} = -17.204 + 1.202(24) \approx 11.6$ crimes per 1,000

17. (a)

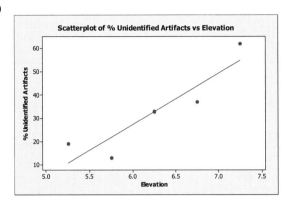

(b) Use a calculator to verify.

(c) $\bar{x} = \dfrac{\sum x}{n} = \dfrac{31.25}{5} = 6.25$

$\bar{y} = \dfrac{\sum y}{n} = \dfrac{164}{5} = 32.8$

$b = \dfrac{n\sum xy - \left(\sum x\right)\left(\sum y\right)}{n\sum x^2 - \left(\sum x\right)^2} = \dfrac{5(1,080) - (31.25)(164)}{5(197.813) - (31.25)^2} = 22.0$

$a = \bar{y} - b\bar{x} = 32.8 - 22.0(6.25) = -104.5$

$\hat{y} = a + bx$ or $\hat{y} = -104.7 + 22.0x$

(d) See the figure in part (a).

(e) $r^2 = (0.913)^2 \approx 0.833$

This means that 83.3% of the variation in y = % unidentified artifacts can be explained by the corresponding variation in x = elevation. $100\% - 83.3\% = 16.7\%$ of the variation is unexplained.

(f) Use $x = 6.5$.

$\hat{y} = -104.7 + 22.0(6.5) = 38.3$ percent

19. (a) Yes. The pattern of residuals appears randomly scattered around the horizontal line at 0.

(b) No. There do not appear to be any outliers.

21. (a) Result checks.

(b) Result checks.

(c) Yes.

(d) $y = 0.066 + 1.393x$

$y - 0.066 = 1.393x$

$\dfrac{y - 0.066}{1.393} = x$

$\dfrac{1}{1.393}y - \dfrac{0.066}{1.393} = x$

or $x = 0.718y - 0.047$

The equation $x = 0.718y - 0.047$ does not match part (b) with the symbols x and y exchanged.

(e) In general, switching x and y values produces a different least-squares equation. It is important that when you perform a linear regression, you know which variable is the explanatory variable and which is the response variable.

23. **(a)**

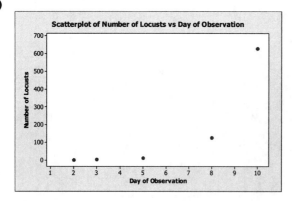

These data are not described well by a straight line because the y values seem to explode as x increases.

(b)

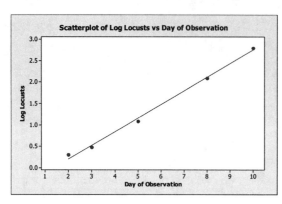

The transformed data are better represented by a straight line.

(c)

$$\hat{y} \approx -0.421 + 0.318x$$
$$r \approx 0.998$$

(d)

$$\alpha \approx 10^{-0.421} \approx 0.379$$
$$\beta \approx 10^{0.318} \approx 2.080$$
$$y \approx 0.379(2.080)^x$$

25. **(a)**

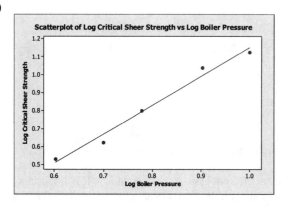

The transformed data seem to be represented well by a straight line.

(b) Using the transformed data,

$\hat{y} \approx -0.451 + 1.600x$

$r \approx 0.991$

(c)

$\alpha \approx 10^{-0.451} \approx 0.351$

$\beta \approx b \approx 1.600$

$\hat{y} \approx 0.354(x)^{1.600}$

Section 9.3

1. The symbol is rho, ρ.

3. As x moves farther from $\bar{x}$, the confidence interval for the predicted y becomes wider.

5. **(a)** The predictor variable is diameter.
   (b) Here, $a = -0.223$, $b = 0.7848$, and $\hat{y} = -0.223 + 0.7848x$.
   (c) The P value of b is 0.001. We are testing H_0: $\beta = 0$ versus H_1: $\beta \neq 0$, and we reject H_0 and conclude that the slope is not 0.
   (d) Here, $r \approx 0.896$. Yes, the correlation coefficient is significant at the $\alpha = 0.01$ level because the P value $< \alpha$.

7. **(a)** Use a calculator to verify.
   (b) $\alpha = 0.05$

   H_0: $\rho = 0$

   H_1: $\rho > 0$

   $t = \dfrac{r\sqrt{n-2}}{\sqrt{1-r^2}} = \dfrac{0.784\sqrt{6-2}}{\sqrt{1-0.784^2}} \approx 2.526$; $\quad d.f. = n - 2 = 6 - 2 = 4$

   From Table 6 in Appendix II, 2.526 falls between entries 2.132 and 2.776. Use the one-tailed areas to find that $0.025 < P$ value < 0.050. Since the P-value interval ≤ 0.05, we reject H_0.

   At the 5% level of significance, there seems to be a positive correlation between x and y.
   (c) Use a calculator to verify.
   (d) $\hat{y} = a + bx$ or $\hat{y} = 16.542 + 0.4117x$

   Use $x = 70$.

   $\hat{y} = 16.542 + 0.4117(70) \approx 45.36\%$

   (e) $d.f. = 4$, $t_c = 2.132$

   $E = t_c S_e \sqrt{1 + \dfrac{1}{n} + \dfrac{n(x-\bar{x})^2}{n\sum x^2 - (\sum x)^2}} = 2.132(2.6964)\sqrt{1 + \dfrac{1}{6} + \dfrac{6(70-73.167)^2}{6(32.393)-(439)^2}} \approx 6.31$

   The 90% confidence interval is

   $\hat{y} - E \leq y \leq \hat{y} + E$

   $45.36 - 6.31 \leq y \leq 4,536 + 6.31$

   $39.05 \leq y \leq 51.67$

(f) $\alpha = 0.05$

$H_0: \beta = 0$

$H_1: \beta > 0$

$$t = \frac{b}{S_e}\sqrt{\sum x^2 - \frac{1}{n}\left(\sum x\right)^2} = \frac{0.4117}{2.6964}\sqrt{32,393 - \frac{1}{6}(439)^2} \approx 2.522 \;; d.f. = n-2 = 6-2 = 4$$

From Table 6 in Appendix II, 2.522 falls between entries 2.132 and 2.776. Use the one-tailed areas to find that $0.025 < P$ value < 0.050. Since the P-value interval < 0.05, we reject H_0.

At the 5% level of significance, there seems to be a positive slope between x and y.

(g) $d.f. = 4$, $t_c = 2.132$, $b \approx 0.4117$

$$E = \frac{t_c S_e}{\sqrt{\sum x^2 - \frac{1}{n}\left(\sum x\right)^2}} \approx \frac{2.132(2.6964)}{\sqrt{32,393 - \frac{1}{6}(439)^2}} \approx 0.3480$$

The 90% confidence interval is

$$b - E < \beta < b + E$$

$$0.4117 - 0.3480 < \beta < 0.4117 + 0.3480$$

$$0.064 < \beta < 0.760$$

For every percentage increase in successful free throws, the percentage of successful field goals increases by an amount between 0.06 and 0.76.

9. **(a)** Use a calculator to verify.

(b) $\alpha = 0.01$

$H_0: \rho = 0$

$H_1: \rho < 0$

$$t = \frac{r\sqrt{n-2}}{\sqrt{1-r^2}} = \frac{-0.976\sqrt{7-2}}{\sqrt{1-(-0.976)^2}} = -10.02\;; \qquad d.f. = n-2 = 7-2 = 5$$

From Table 6 in Appendix II, 10.02 falls to the right of entry 6.869. Use the one-tailed areas to find P value < 0.0005. Using a TI-84, P value ≈ 0.00008.

Since the P-value interval ≤ 0.01, we reject H_0.

At the 1% level of significance, the evidence supports a negative correlation between x and y.

(c) Use a calculator to verify.

(d) $\hat{y} = a + bx$ or $\hat{y} = 3.366 - 0.0544x$

Use $x = 18$.

$\hat{y} = 3.366 - 0.0544(18) \approx 2.39$ hours

(e) $d.f. = 5$, $t_c = 1.476$, $\bar{x} = \dfrac{\sum x}{n} = \dfrac{201.4}{7} \approx 28.77$

$$E = t_c S_e \sqrt{1 + \frac{1}{n} + \frac{n(x-\bar{x})^2}{n\sum x^2 - \left(\sum x\right)^2}} = 1.476(0.1660)\sqrt{1 + \frac{1}{7} + \frac{7(18-28.77)^2}{7(6,735.46) - (201.4)^2}} \approx 0.276$$

The 80% confidence interval is
$$\hat{y} - E \le y \le \hat{y} + E$$
$$2.39 - 0.276 \le y \le 2.39 + 0.276$$
$$2.11 \le y \le 2.67$$

(f) $\alpha = 0.01$

$H_0: \beta = 0$

$H_1: \beta < 0$

$$t = \frac{b}{S_e}\sqrt{\sum x^2 - \frac{1}{n}\left(\sum x\right)^2} = \frac{-0.0544}{0.1660}\sqrt{6,735.46 - \frac{1}{7}(201.4)^2} \approx -10.02$$

From Table 6 in Appendix II, 10.02 falls to the right of entry 6.869. Use the one-tailed areas to find P value < 0.0005. Using a TI-84, P value ≈ 0.00008.
Since the P-value interval ≤ 0.01, we reject H_0.
At the 1% level of significance, the sample evidence supports a negative slope.

(g) $d.f. = 5$, $t_c = 2.015$, $b \approx -0.0544$

$$E = \frac{t_c S_e}{\sqrt{\sum x^2 - \frac{1}{n}\left(\sum x\right)^2}} = \frac{2.015(0.1660)}{\sqrt{6,735.46 - \frac{1}{7}(201.4)^2}} \approx 0.011$$

The 90% confidence interval is
$$b - E < \beta < b + E$$
$$-0.054 - 0.011 < \beta < -0.054 + 0.011$$
$$-0.065 < \beta < -0.043$$

For a 1-m increase in depth, the optimal time decreases from about 0.04 to 0.07 hour.

11. (a) Use a calculator to verify.

(b) $\alpha = 0.01$

$H_0: \rho = 0$

$H_1: \rho > 0$

$$t = \frac{r\sqrt{n-2}}{\sqrt{1-r^2}} = \frac{0.956\sqrt{6-2}}{\sqrt{1-(0.956)^2}} = 6.517 \; ; d.f. = n - 2 = 6 - 2 = 4$$

From Table 6 in Appendix II, the sample test statistic falls between entries 4.604 and 8.610. Use one-tailed areas to find that $0.0005 < P$ value < 0.005. Using a TI-84, P value ≈ 0.0014.
Since the P-value interval ≤ 0.01, reject H_0.
At the 1% level of significance, the sample evidence supports a positive correlation.

(c) Use a calculator to verify.

(d) $\hat{y} = a + bx$ or $\hat{y} = 1.965 + 0.7577x$

Use $x = 14$.

$\hat{y} = 1.965 + 0.7577(14) \approx \$12.57$ thousand.

(e) $d.f. = 4$, $t_c = 1.778$, $\bar{x} = \frac{\sum x}{n} = \frac{79.6}{6} \approx 13.267$

$$E = t_c S_e \sqrt{1 + \frac{1}{n} + \frac{n(x-\bar{x})^2}{n\sum x^2 - \left(\sum x\right)^2}} = 1.778(0.1527)\sqrt{1 + \frac{1}{6} + \frac{6(14-13.267)^2}{6(1,057.8)-(79.6)^2}} \approx 0.329$$

The 85% confidence interval is
$$\hat{y} - E \le y \le \hat{y} + E$$
$$12.57 - 0.329 \le y \le 12.57 + 0.329$$
$$12.24 \le y \le 12.90$$

(f) $\alpha = 0.01$
$H_0: \beta = 0$
$H_1: \beta > 0$

$$t = \frac{b}{S_e}\sqrt{\sum x^2 - \frac{1}{n}\left(\sum x\right)^2} = \frac{0.7577}{0.1527}\sqrt{(1{,}057.8) - \frac{1}{6}(79.6)^2} \approx 6.608; \quad d.f. = n - 2 = 6 - 2 = 4$$

From Table 6 in Appendix II, the sample test statistic falls between entries 4.604 and 8.610. Use one-tailed areas to find that $0.0005 < P$ value < 0.005. Using a TI-84, P value ≈ 0.0014.
Since the P-value interval ≤ 0.01, reject H_0.
At the 1% level of significance, the sample evidence supports a positive slope.

(g) $d.f. = 4$, $t_c = 2.776$, $b \approx 0.7577$

$$E = \frac{t_c S_e}{\sqrt{\sum x^2 - \frac{1}{n}\left(\sum x\right)^2}} = \frac{2.776(0.1527)}{\sqrt{(1{,}057.8) - \frac{(79.6)^2}{6}}} \approx 0.318$$

The 95% confidence interval is
$$b - E < \beta < b + E$$
$$0.758 - 0.318 < \beta < 0.758 + 0.318$$
$$0.44 < \beta < 1.08$$

For every \$1,000 increase in list price, there is an increase in dealer price of between \$440 and \$1,080.

13. (a) $\alpha = 0.01$
$H_0: \rho = 0$
$H_1: \rho \ne 0$

$$t = \frac{r\sqrt{n-2}}{\sqrt{1-r^2}} = \frac{0.90\sqrt{6-2}}{\sqrt{1-(0.90)^2}} = 4.129$$
$$d.f. = n - 2 = 6 - 2 = 4$$

From Table 6 in Appendix II, 4.129 falls between entries 3.747 and 4.604. Use two-tailed areas to find $0.010 < P$ value < 0.020. Since the P-value interval > 0.01, we do not reject H_0. The correlation coefficient ρ is not significantly different from 0 at the 0.01 level of significance.

(b) $\alpha = 0.01$
$H_0: \rho = 0$
$H_1: \rho \ne 0$

$$t = \frac{r\sqrt{n-2}}{\sqrt{1-r^2}} = \frac{0.90\sqrt{10-2}}{\sqrt{1-(0.90)^2}} = 5.840$$
$$d.f. = n - 2 = 10 - 2 = 8$$

From Table 6 in Appendix II, 5.840 falls to the right of entry 5.041. Use two-tailed areas to find P value < 0.001. Since the P value ≤ 0.01, we reject H_0. The correlation coefficient ρ is significantly different from 0 at the 0.01 level of significance.

(c) From part (a) to part (b), n increased from 6 to 10, and the test statistic t increased from 4.129 to 5.840. For the same $r = 0.90$ and the same level of significance $\alpha = 0.01$, we rejected H_0 for the larger n but not for the smaller n.

In general, as n increases, the degrees of freedom $(n - 2)$ increase, and the test statistic $\left(t = \dfrac{r\sqrt{n-2}}{\sqrt{1-r^2}} \right)$

increases. This produces a smaller P value.

15. (a) Use a calculator to verify:
$\hat{y} = a + bx$ or $\hat{y} \approx 1.9938 + 0.9165x$
Use $x = 5.8$.
$\hat{y} \approx 1.9938 + 0.9165(5.8) \approx 7.3095$

(b) Use a calculator to verify $r \approx 0.9815$ and $r^2 \approx 0.9633$
$H_0: \rho = 0$
$H_1: \rho > 0$
$t = \dfrac{r\sqrt{n-2}}{\sqrt{1-r^2}} = \dfrac{0.9815\sqrt{7-2}}{\sqrt{1-(0.9815)^2}} \approx 11.4628$
$d.f. = n - 2 = 7 - 2 = 5$

Using a TI-84, P value ≈ 0.000044. Since the P-value ≤ 0.01, reject H_0. The data support a positive correlation and indicate a predictable original time series from one week to the next.

17. (b) Use a calculator to verify:
$\hat{y} = a + bx$ or $\hat{y} \approx 4.1415 + 0.9785x$
Use $x = \$42$.
$\hat{y} \approx 4.1415 + 0.9785(42) \approx \$45.24$

(c) Use a calculator to verify $r \approx 0.9668$ and $r^2 \approx 0.9347$
$H_0: \rho = 0$
$H_1: \rho > 0$
$t = \dfrac{r\sqrt{n-2}}{\sqrt{1-r^2}} = \dfrac{0.9668\sqrt{6-2}}{\sqrt{1-(0.9668)^2}} \approx 7.5669$
$d.f. = n - 2 = 6 - 2 = 4$

Using a TI-84, P value ≈ 0.0008. Since the P-value ≤ 0.01, reject H_0. The data support a positive correlation and indicate a predictable original time series from one year to the next.

Section 9.4

1. $x_1 = 1.6 + 3.5x_2 - 7.9x_3 + 2.0x_4$

(a) The response variable is x_1. The explanatory variables are x_2, x_3, and x_4.

(b) The constant term is 1.6.
The coefficient 3.5 corresponds to x_2.
The coefficient -7.9 corresponds to x_3.
The coefficient 2.0 corresponds to x_4.

(c) $x_2 = 2, x_3 = 1, x_4 = 5$

$x_1 = 1.6 + 3.5(2) - 7.9(1) + 2.0(5) = 10.7$

The predicted value is 10.7.

(d) In multiple regression, the coefficients of the explanatory variables can be thought of as "slopes" if we look at one explanatory variable's coefficient at a time while holding the other explanatory variables as arbitrary and fixed constants.

x_3 and x_4 held constant, x_2 increased by one unit, the change in x_1 would be an increase of 3.5 units.

x_3 and x_4 held constant, x_2 increased by two units, the change in x_1 would be an increase of 7 units.

x_3 and x_4 held constant, x_2 decreased by four units, the change in x_1 would be a decrease of 14 units.

(e) $d.f. = n - k - 1 = 12 - 3 - 1 = 8$

A 90% confidence interval for the coefficient of x_2 is $b_2 - tS_2 < \beta_2 < b_2 + tS_2$.

$3.5 - 1.86(0.419) < \beta_2 < 3.5 + 1.86(0.419)$

$2.72 < \beta_2 < 4.28$

(f) $\alpha = 0.05$

$H_0: \beta_2 = 0$ $d.f. = 8$

$H_1: \beta_2 \neq 0$

$t = \dfrac{b_2 - \beta_2}{S_2} = \dfrac{3.5 - 0}{0.419} = 8.35$

From Table 6 in Appendix II, 8.35 falls to the right of entry 5.041. Use two-tailed areas to find that P-value < 0.001. Since $0.001 \leq 0.05$, reject H_0.

We conclude that $\beta_2 \neq 0$ and x_2 should be included as an explanatory variable in the least-squares equation.

3. (a)

	$\bar{x}$	s	$CV = \dfrac{s}{\bar{x}} \cdot 100$
x_1	150.09	13.63	9.08%
x_2	62.45	9.11	14.59%
x_3	195.0	17.31	8.88%

Relative to its mean, x_2 has the greatest spread of data values, and x_3 has the smallest spread of data values.

(b) $r^2_{x_1x_2} \approx (0.979)^2 \approx 0.958$

$r^2_{x_1x_3} \approx (0.971)^2 \approx 0.943$

$r^2_{x_2x_3} \approx (0.946)^2 \approx 0.895$

The variable x_2 has the greatest influence on x_1 $(0.958 > 0.943)$.

Yes. Both variables x_2 and x_3 show a strong influence on x_1 because 0.958 and 0.943 are close to 1.

95.8% of the variation of x_1 can be explained by the corresponding variation in x_2.

94.3% of the variation of x_1 can be explained by the corresponding variation in x_3.

(c) $R^2 = 0.977$

97.7% of the variation in x_1 can be explained by the corresponding variation in x_2 and x_3 taken together.

(d) $x_1 = 30.99 + 0.861x_2 + 0.355x_3$

In multiple regression, the coefficients of the explanatory variables can be thought of as "slopes" if we look at one explanatory variable's coefficient at a time while holding the other explanatory variables as arbitrary and fixed constants.

If age (x_2) were held fixed and x_3 increased by 10 pounds, the systolic blood pressure would be expected to increase by $0.335(10) = 3.35$.
If weight (x_3) were held fixed and x_2 increased by 10 years, the systolic blood pressure would be expected to increase by $0.861(10) = 8.61$.

(e) $\alpha = 0.05$

$H_0: \beta_i = 0$
$H_1: \beta_i \neq 0$
$d.f = n - k - 1 = 11 - 2 - 1 = 8$

For β_2, the sample test statistic is $t = 3.47$ with P value $= 0.008$.
For β_3, the sample test statistic is $t = 2.56$ with P value $= 0.034$.
Since $0.008 \leq 0.05$ and $0.034 \leq 0.05$, reject H_0 for each coefficient and conclude that the coefficients of x_2 and x_3 are not zero. Explanatory variables x_i whose coefficients (β_i) are nonzero, contribute information in the least-squares equation; i.e., without these x_i, the resulting least-squares regression equation is not as good a fit to the data as is the regression equation that includes these x_i.

(f) $d.f. = 8, t = 1.86$
A 90% confidence interval for β_i is
$$b_i - tS_i < \beta_i < b_i + tS_i$$
$$0.861 - 1.86(0.2482) < \beta_2 < 0.861 + 1.86(0.2482)$$
$$0.40 < \beta_2 < 1.32$$
$$0.335 - 1.86(0.1307) < \beta_3 < 0.335 + 1.86(0.1307)$$
$$0.09 < \beta_3 < 0.58$$

(g) $x_1 = 30.99 + 0.861(68) + 0.335(192) \approx 153.9$
Michael's predicted systolic blood pressure is 153.9, and a 90% confidence interval for this new observation's value, given these x_i (i.e., the prediction interval), is 148.3 to 159.4.

5. (a)

	$\bar{x}$	s	$CV = \dfrac{s}{\bar{x}} \cdot 100$
x_1	85.24	33.79	39.64%
x_2	8.74	3.89	44.51%
x_3	4.90	2.48	50.61%
x_4	9.92	5.17	52.12%

Relative to its mean, x_4 has the largest spread of data values. The larger the CV, the more we expect the variable to change relative to its average value because a variable with a large CV has a large standard deviation s relative to $\bar{x}$, and s measures "spread," or variability, in the data. x_1 has a small CV because we divide by a large mean.

(b) $r^2_{x_1x_2} \approx (0.917)^2 \approx 0.841$

$r^2_{x_1x_3} \approx (0.930)^2 \approx 0.865$

$r^2_{x_1x_4} \approx (0.475)^2 \approx 0.226$

$r^2_{x_2x_3} \approx (0.790)^2 \approx 0.624$

$r^2_{x_2x_4} \approx (0.429)^2 \approx 0.184$

$r^2_{x_3x_4} \approx (0.299)^2 \approx 0.089$

The variable x_4 has the least influence on box office receipts x_1 ($0.226 < 0.841 < 0.865$).

x_2 = production costs, $r^2_{x_1x_2} \approx 0.841$.

84.1% of the variation of box office receipts can be attributed to the corresponding variation in production costs.

(c) $R^2 = 0.967$

96.7% of the variation in x_1 can be explained by the corresponding variation in x_2, x_3, and x_4 taken together.

(d) $x_1 = 7.68 + 3.66x_2 + 7.62x_3 + 0.83x_4$

In multiple regression, the coefficients of the explanatory variables can be thought of as "slopes" if we look at one explanatory variable's coefficient at a time while holding the other explanatory variables as arbitrary and fixed constants.

If x_2 and x_4 were held fixed and x_3 were increased by 1 ($1 million), the corresponding change in x_1 (box office receipts) would be an increase of $7.62 million.

(e) $\alpha = 0.05$

$H_0: \beta_i = 0$

$H_1: \beta_i \neq 0$

$d.f = n - k - 1 = 10 - 3 - 1 = 6$

For β_2, the sample test statistic is $t = 3.28$ with P value $= 0.017$.

For β_3, the sample test statistic is $t = 4.60$ with P value $= 0.004$.

For β_4, the sample test statistic is $t = 1.54$ with P value $= 0.175$.

Since $0.017 \leq 0.05$ and $0.004 \leq 0.05$, reject H_0 for coefficients β_2 and β_3 and conclude that the coefficients of x_2 and x_3 are not zero.

Since $0.175 > 0.05$, do not reject H_0 for the coefficient β_4 and conclude that the coefficient of x_4 could be zero. If $\beta_4 = 0$, then x_4 contributes nothing to the population regression line. We can eliminate the variable x_4 and fit the estimated regression line without it and probably see little, if any, difference between the predicted values of x_1 based on x_2 and x_3 only and the predicted values of x_1 based on x_2, x_3, and x_4.

(f) $d.f. = 6$, $t = 1.943$

A 90% confidence interval for β_i is

$$b_i - tS_i < \beta_i < b_i + tS_i$$

$$3.662 - 1.943(1.118) < \beta_2 < 3.662 + 1.943(1.118)$$
$$1.49 < \beta_2 < 5.83$$

$$7.621 - 1.943(1.657) < \beta_3 < 7.621 + 1.943(1.657)$$
$$4.40 < \beta_3 < 10.84$$

$$0.8285 - 1.943(0.5394) < \beta_4 < 0.8285 + 1.943(0.5394)$$
$$-0.22 < \beta_4 < 1.88$$

(g) $x_1 = 7.68 + 3.66(11.4) + 7.62(4.7) + 0.83(8.1) = 91.94$

The prediction is \$91.94 million, and a 85% confidence interval for this new observation's value, given these x_i (i.e., the prediction interval), is \$77.6 million to \$106.3 million.

(h) $x_3 = -0.650 + 0.102x_1 - 0.260x_2 - 0.0899x_4$

$x_3 = -0.650 + 0.102(100) - 0.260(12) - 0.0899(9.2)$

$x_3 = 5.63$

The prediction is \$5.63 million, and a 80% confidence interval for this new observation's value, given these x_i (i.e. the prediction interval), is \$4.21 million to \$7.04 million.

7. Answers vary.

Chapter Review Problems

1. We expect r to be close to 0.

3. Results are more reliable for interpolation.

5. **(a)**

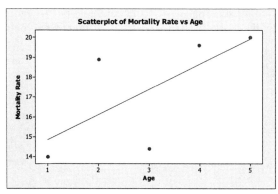

(b) $\bar{x} = \dfrac{\sum x}{n} = \dfrac{15}{5} = 3$

$\bar{y} = \dfrac{\sum y}{n} = \dfrac{86.9}{5} = 17.38$

$b = \dfrac{n\sum xy - \left(\sum x\right)\left(\sum y\right)}{n\sum x^2 - \left(\sum x\right)^2} = \dfrac{5(273.4) - (15)(86.9)}{5(55) - (15)^2} = 1.27$

$a = \bar{y} - b\bar{x} = 17.38 - 1.27(3) = 13.57$

$y = a + bx$ or $y = 13.57 + 1.27x$

(c) $r = \dfrac{n\sum xy - \left(\sum x\right)\left(\sum y\right)}{\sqrt{n\sum x^2 - \left(\sum x\right)^2}\ \sqrt{n\sum y^2 - \left(\sum y\right)^2}} = \dfrac{5(273.4) - (15)(86.9)}{\sqrt{5(55) - (15)^2}\ \sqrt{5(1,544.73) - (86.9)^2}} \approx 0.685$

$r^2 = (0.685)^2 \approx 0.469$

The correlation coefficient r measures the strength of the linear relationship between a bighorn sheep's age and the mortality rate. The coefficient of determination r^2 means that 46.9% of the variation in mortality rate can be explained by the corresponding variation in age of a bighorn sheep using the least-squares line.

(d) $\alpha = 0.01$

$H_0: \rho = 0$ $d.f. = n - 2 = 5 - 2 = 3$

$H_1: \rho > 0$

$t = \dfrac{r\sqrt{n-2}}{\sqrt{1-r^2}} = \dfrac{0.685\sqrt{5-2}}{\sqrt{1-(0.685)^2}} = 1.629$

From Table 6 in Appendix II, 1.629 falls between entries 1.423 and 1.638. Use one-tailed areas to find that $0.100 < P$ value < 0.125. Using a TI-84, P value ≈ 0.1011.
Since the P-value interval is > 0.01, we do not reject H_0.

There does not seem to be a positive correlation between age and mortality rate of bighorn sheep.

(e)   No, based on this limited data, predictions from the least-squares line model might be misleading. There appear to be other lurking variables that affect the mortality rate of sheep in different age groups.

7.   **(a)**

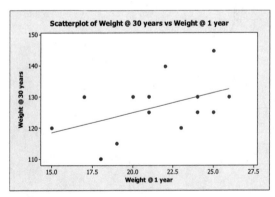

(b) $\bar{x} = \dfrac{\sum x}{n} = \dfrac{300}{14} = 21.43$

$\bar{y} = \dfrac{\sum y}{n} = \dfrac{1,775}{14} = 126.79$

$b = \dfrac{n\sum xy - (\sum x)(\sum y)}{n\sum x^2 - (\sum x)^2} = \dfrac{14(38,220) - (300)(1,775)}{14(6,572) - (300)^2} \approx 1.285$

$a = \bar{y} - b\bar{x} = 126.79 - (1.285)(21.43) = 99.25$

$\hat{y} = a + bx$ or $\hat{y} = 99.25 + 1.285x$

(c) $r = \dfrac{n\sum xy - (\sum x)(\sum y)}{\sqrt{n\sum x^2 - (\sum x)^2}\sqrt{n\sum y^2 - (\sum y)^2}} = \dfrac{14(38,220) - (300)(1,775)}{\sqrt{14(6,572) - (300)^2}\sqrt{14(226,125) - (1,775)^2}} \approx 0.468$

$r^2 = (0.468)^2 \approx 0.219$

The coefficient of determination r^2 means that 21.9% of the variation in weight of a 30-year-old female can be explained by the corresponding variation in the weight of a 1-year-old baby girl using the least-squares line.

(d) $\alpha = 0.01$

$H_0: \rho = 0$ $d.f. = n - 2 = 14 - 2 = 12$

$H_1: \rho > 0$

$t = \dfrac{r\sqrt{n-2}}{\sqrt{1-r^2}} = \dfrac{0.468\sqrt{14-2}}{\sqrt{1-(0.468)^2}} = 1.834$

From Table 6 in Appendix II, the sample test statistic falls between entries 1.782 and 2.179. Use one-tailed areas to find that $0.025 < P$ value < 0.050. Using a TI-84, P value ≈ 0.0457. Since the P-value interval > 0.01, do not reject H_0. At the 1% level of significance, there does not seem to be a positive correlation between weight of baby and weight of adult.

(e) Let $x = 20$.

$\hat{y} = 99.25 + 1.285(20) = 124.95$

The predicted weight is 124.95 pounds, but since r is not significant, the prediction might not be very reliable.

(f) $S_e = \sqrt{\dfrac{\sum y^2 - a\sum y - b\sum xy}{n-2}} = \sqrt{\dfrac{226,125 - 99.25(1,775) - 1.285(38,220)}{14-2}} \approx 8.38$

(g) $E = t_c S_e \sqrt{1 + \dfrac{1}{n} + \dfrac{n(x-\bar{x})^2}{n\sum x^2 - (\sum x)^2}}$

$= 2.179(8.38)\sqrt{1 + \dfrac{1}{14} + \dfrac{14(20-21.43)^2}{14(6,572) - (300)^2}}$

≈ 19.03

A 95% confidence interval for y is

$\hat{y} - E \le y \le \hat{y} + E$

$124.95 - 19.03 \le y \le 124.95 + 19.03$

$105.92 \le y \le 143.98$

(h) $\alpha = 0.01$

$H_0: \beta = 0$ $\qquad\qquad\qquad\qquad d.f. = n - 2 = 14 - 2 = 12$

$H_1: \beta > 0$

$$t = \frac{b}{S_e}\sqrt{\sum x^2 - \frac{1}{n}\left(\sum x\right)^2} = \frac{1.285}{8.38}\sqrt{6,572 - \frac{1}{14}(300)^2} \approx 1.84$$

From Table 6 in Appendix II, the sample test statistic falls between entries 1.782 and 2.179. Use one-tailed areas to find that $0.025 < P$ value < 0.050. Using a TI-84, P value ≈ 0.0457. Since the P-value interval > 0.01 for α, do not reject H_0. At the 1% level of significance, there does not seem to be a positive slope between weight of baby and weight of adult.

(i) $d.f. = 12,\ t_c = 1.356,\ b = 1.285$

$$E = \frac{t_c S_e}{\sqrt{\sum x^2 - \frac{(\sum x)^2}{n}}} = \frac{1.356(8.38)}{\sqrt{6,572 - \frac{(300)^2}{14}}} \approx 0.949$$

An 80% confidence interval is

$$b - E < \beta < b + E$$

$$1.285 - 0.949 < \beta < 1.285 + 0.949$$

$$0.34 < \beta < 2.22$$

At the 80% confidence level, we can say that for each additional pound a female infant weighs at 1 year, the adult weight increases by 0.34–2.22 pounds.

9. (a)

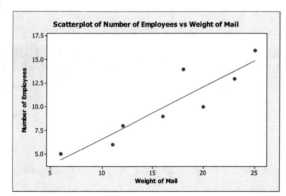

(b) $\bar{x} = \frac{\sum x}{n} = \frac{131}{8} = 16.375 \approx 16.38$

$\bar{y} = \frac{\sum y}{n} = \frac{81}{8} = 10.125 \approx 10.13$

$b = \frac{n\sum xy - (\sum x)(\sum y)}{n\sum x^2 - (\sum x)^2} = \frac{8(1,487) - (131)(81)}{8(2,435) - (131)^2} \approx 0.554118 \approx 0.554$

$a = \bar{y} - b\bar{x} = 10.125 - 0.554118(16.375) = 1.051$

$\hat{y} = a + bx$ or $\hat{y} = 1.051 + 0.554x$

(c) $r = \frac{n\sum xy - (\sum x)(\sum y)}{\sqrt{n\sum x^2 - (\sum x)^2}\sqrt{n\sum y^2 - (\sum y)^2}} = \frac{8(1,487) - (131)(81)}{\sqrt{8(2,435) - (131)^2}\sqrt{8(927) - (81)^2}} \approx 0.913$

$r^2 = (0.913)^2 \approx 0.833$

The coefficient of determination r^2 means that 83.3% of the variation in number of employees can be explained by the corresponding variation in weight of incoming mail using the least-squares line.

(d) $\alpha = 0.01$

$H_0: \rho = 0$ $d.f. = n - 2 = 8 - 2 = 6$

$H_1: \rho > 0$

$t = \dfrac{r\sqrt{n-2}}{\sqrt{1-r^2}} = \dfrac{0.913\sqrt{8-2}}{\sqrt{1-(0.913)^2}} \approx 5.48$

From Table 6 in Appendix II, the sample test statistic falls between entries 3.707 and 5.959. Use one-tailed areas to find that $0.0005 < P$ value < 0.005. Using a TI-84, P value ≈ 0.0008. Since the P-value interval ≤ 0.01, reject H_0. At the 1% level of significance, there is sufficient evidence to show a positive correlation between pounds of mail and number of employees required to process the mail.

(e)   Use $x = 15$.

$\hat{y} = 1.051 + 0.554(15) = 9.36$

About nine employees should be assigned mail duty.

(f) $S_e = \sqrt{\dfrac{\sum y^2 - a\sum y - b\sum xy}{n-2}} = \sqrt{\dfrac{927 - 1.051(81) - 0.554118(1,487)}{8-2}} \approx 1.73$

(g) $E = t_c S_e \sqrt{1 + \dfrac{1}{n} + \dfrac{n(x-\overline{x})^2}{n\sum x^2 - \left(\sum x\right)^2}}$

$= 2.447(1.73)\sqrt{1 + \dfrac{1}{8} + \dfrac{8(15-16.375)^2}{8(2,435)-(131)^2}}$

≈ 4.5

A 95% confidence interval for y is

$\hat{y} - E \leq y \leq \hat{y} + E$

$9.36 - 4.5 \leq y \leq 9.36 + 4.5$

$4.86 \leq y \leq 13.86$

(h) $\alpha = 0.01$

$H_0: \beta = 0$ $d.f. = n - 2 = 8 - 2 = 6$

$H_1: \beta > 0$

$t = \dfrac{b}{S_e}\sqrt{\sum x^2 - \dfrac{1}{n}\left(\sum x\right)^2} = \dfrac{0.554118}{1.73}\sqrt{2435 - \dfrac{1}{8}(131)^2} \approx 5.45$

From Table 6 in Appendix II, the sample test statistic falls between entries 3.707 and 5.959. Use one-tailed areas to find that $0.0005 < P$ value < 0.005. Using a TI-84, P value ≈ 0.0008. Since the P-value interval ≤ 0.01, reject H_0. At the 1% level of significance, there is sufficient evidence to show a positive slope between pounds of mail and number of employees required to process the mail.

(i) $d.f. = 6, t_c = 1.440, b \approx 0.554$

$$E = \frac{t_c S_e}{\sqrt{\sum x^2 - \frac{(\sum x)^2}{n}}} = \frac{1.440(1.73)}{\sqrt{2,435 - \frac{(131)^2}{8}}} \approx 0.146$$

An 80% confidence interval is
$$b - E < \beta < b + E$$
$$0.554 - 0.146 < \beta < 0.554 + 0.146$$
$$0.41 < \beta < 0.70$$

At the 80% confidence level, we can say that for each additional pound of mail, between 0.4 and 0.7 additional employees are needed.

Cumulative Review Problems Chapters 7, 8, 9

1. **(a)** H_0: $\mu = 2.0$
    H_1: $\mu > 2.0$
    $\alpha = 0.01$
    Since σ is known and x is normal, we will use the standard normal distribution.
    $$z = \frac{2.56 - 2.0}{0.7 \big/ \sqrt{10}} = 2.53$$
    The P value will be $P(z > 2.53) = 0.0057$.

    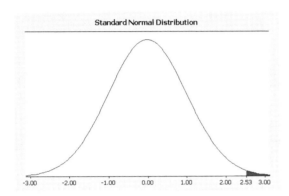

    Standard Normal Distribution

    Since the P value $= 0.0057 < 0.01$, we reject H_0. Yes, the data are statistically significant.
    At the 1% level of significance, the evidence is sufficient to say that the population mean discharge level of lead is higher.

    (b)
    $$\overline{x} - z \times \sigma \big/ \sqrt{n} < \mu < \overline{x} + z \times \sigma \big/ \sqrt{n}$$
    $$2.56 - 1.96 \times 0.7 \big/ \sqrt{10} < \mu < 2.56 + 1.96 \times 0.7 \big/ \sqrt{10}$$
    $$2.13 < \mu < 2.99$$

    (c) $n \approx \left(\dfrac{z_c \sigma}{E} \right)^2 = \left(\dfrac{1.96 \times 0.7}{0.2} \right)^2 = 47.06$, so use $n = 48$ water samples.

3. **(a)** Check conditions: $np = 68 \times 0.10 = 6.8$, $nq = 68 \times 0.90 = 61.2$. Since both estimates are reasonably greater than 5, the conditions are satisfied.
    H_0: $p = 0.10$
    H_1: $p \neq 0.10$
    $\alpha = 0.05$
    Since the conditions are satisfied, $\hat{p}$ follows a normal distribution.

$$z = \frac{0.147 - 0.10}{\sqrt{\dfrac{0.10 \times 0.90}{68}}} = 1.29$$

The P value will be $2 \times P(z > 1.29) = 0.197$.

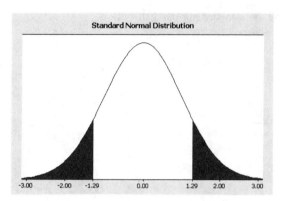

Since the P value > 0.05, we fail to reject H_0. No, the data are not statistically significant.
At the 5% level of significance, the data do not indicate any difference from the national average for the population proportion of crime victims.

(b) For a 95% confidence interval, $z_c = 1.96$.

$$\hat{p} - z \times \sqrt{\frac{\hat{p}\hat{q}}{n}} < p < \hat{p} + z \times \sqrt{\frac{\hat{p}\hat{q}}{n}}$$

$$0.147 - 1.96 \times \sqrt{\frac{0.147 \times 0.853}{68}} < p < 0.147 + 1.96 \times \sqrt{\frac{0.147 \times 0.853}{68}}$$

$$0.063 < p < 0.231$$

(c) $n \approx \left(\dfrac{1.96}{0.05}\right)^2 \times 0.147 \times 0.853 \approx 192.68$, so use $n = 193$ students.

5. **(a)** $H_0: \mu_1 = \mu_2$
$H_1: \mu_1 \neq \mu_2$
$\alpha = 0.05$
We will use the Student's t distribution with $d.f. = 16 - 1 = 15$.

$$t = \frac{(74.8 - 70.1) - (0 - 0)}{\sqrt{\left(\dfrac{5.2^2}{16}\right) + \left(\dfrac{8.6^2}{18}\right)}} = 1.952$$

Using the TI-84, the P value will be $2 \times P(t > 1.952) = 0.070$.
Using Table 6 in Appendix II, $0.050 < P$ value < 0.100.

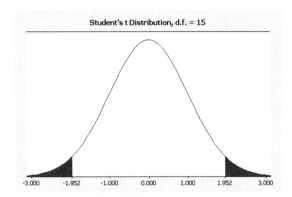

Since the *P*-value interval > 0.05, we fail to reject H_0. No, the data are not statistically significant. At the 5% level of significance, the evidence does not show any difference in the population mean proportion of on-time arrivals in summer versus winter.

(b)

$$\left(\overline{x}_1 - \overline{x}_2\right) - t \times \sqrt{\frac{s_1^{\,2}}{n_1} + \frac{s_2^{\,2}}{n_2}} < \mu_1 - \mu_2 < \left(\overline{x}_1 - \overline{x}_2\right) + t \times \sqrt{\frac{s_1^{\,2}}{n_1} + \frac{s_2^{\,2}}{n_2}}$$

$$\left(74.8 - 70.1\right) - 2.131 \times \sqrt{\frac{5.2^2}{16} + \frac{8.6^2}{18}} < \mu_1 - \mu_2 < \left(74.8 - 70.1\right) + 2.131 \times \sqrt{\frac{5.2^2}{16} + \frac{8.6^2}{18}}$$

$$-0.43\% < \mu_1 - \mu_2 < 9.83\%$$

(c)   We assume that x_1 and x_2 are approximately normal or at least mound-shaped and symmetric.

7.   **(a)**   Essay. Refer to Chapters 7 and 8.
  (b)   Answers will vary.

9.   **(a)**

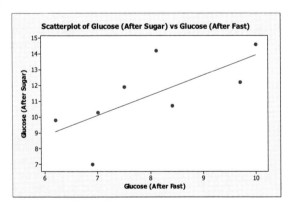

(b) $b = \dfrac{(8)(739.65) - (63.8)(90.7)}{(8)(521.56) - (63.8)^2} \approx 1.279$

$$a = 11.338 - (1.279)(7.975) \approx 1.138$$

$$\hat{y} = 1.138 + 1.279x$$

(c) $r = \dfrac{n\sum xy - \left(\sum x\right)\left(\sum y\right)}{\sqrt{n\sum x^2 - \left(\sum x\right)^2}\,\sqrt{n\sum y^2 - \left(\sum y\right)^2}} = \dfrac{8(739.65) - (63.8)(90.7)}{\sqrt{8(1521.56) - (63.8)^2}\,\sqrt{8(1070.87) - (90.7)^2}} \approx 0.700$

$r^2 = 0.490$

Therefore, 49% of the variance in glucose after drinking sugar water is explained by the model and the variance in glucose after fasting.

(d) $\hat{y} = 1.138 + 1.279(9.0) = 12.65$

$$E = t_c S_e \sqrt{1 + \frac{1}{n} + \frac{n\left(x - \bar{x}\right)^2}{n\sum x^2 - \left(\sum x\right)^2}}$$

$$= 1.440 \times 1.899 \sqrt{1 + \frac{1}{8} + \frac{8(9 - 7.975)^2}{8(521.56) - (63.8)^2}} = 3.01$$

Therefore, an 80% confidence interval for $\hat{y}$ will be

$12.65 - 3.01 < \hat{y} < 12.65 + 3.01$

$9.64 < \hat{y} < 15.66.$

(e) $H_0 : \rho = 0$

$H_1 : \rho \neq 0$

$\alpha = 0.01$

$t = \dfrac{0.700\sqrt{8 - 2}}{\sqrt{1 - 0.700^2}} \approx 2.401$

Since *d.f. = 6, 0.05 < P-value < 0.10*; do not reject H_0. At the 1% level of significance, there is not sufficient evidence to conclude that there is a linear correlation between the two glucose measurements.

(f) $E = \dfrac{t_c S_e}{\sqrt{\sum x^2 - \frac{1}{n}\left(\sum x\right)^2}} = \dfrac{1.650 \times 1.899}{\sqrt{521.56 - \frac{1}{8}(63.8)^2}} \approx 0.88$

Therefore, an 85% confidence interval for β will be

$1.279 - 0.88 < \beta < 1.279 + 0.88$

$0.399 < \beta < 2.159.$

Chapter 10: Chi-Square and F Distributions

Section 10.1

1. Chi-square distributions are skewed right.

3. Use a right-tailed test.

5. Take random samples from each of the four ages groups and record the number of people in each age group who recycle each of the three product types. Make a contingency table with age groups as labels for the rows and product groups as labels for the columns (or vice versa).

7. **(a)** $df = (R-1)(C-1) = (4-1)(3-1) = 6$; Using Table 7, $0.005 < P - \text{value} < 0.010$

    Since the $P - \text{value} < \alpha = 0.01$, reject the null hypothesis. At the 1% level of significance, we conclude that the age groups differ in the proportions of who recycles each of the specified products.

    (b) No. All Zane can say is that the four age groups differ in the proportions of who recycles each specified product. From this study, he cannot determine how the age groups differ.

9. **(a)** $\alpha = 0.05$

    H_0: Myers-Briggs preference and profession are independent.

    H_1: Myers-Briggs preference and profession are not independent.

    (b) $\chi^2 = \sum \dfrac{(O-E)^2}{E}$

    $$= \frac{(62-49.02)^2}{49.02} + \frac{(45-57.98)^2}{57.98} + \frac{(68-74.22)^2}{74.22} + \frac{(94-87.78)^2}{87.78} + \frac{(56-62.76)^2}{62.76} + \frac{(81-74.24)^2}{74.24}$$

    $$= 8.649$$

    All expected frequencies are greater than 5. Use the chi-square distribution.

    Since there are 3 rows and 2 columns, $d.f. = (3-1)(2-1) = 2$.

    (c) In Table 7 in Appendix II, with $d.f. = 2$, $\chi^2 = 8.649$ falls between entries 7.38 and 9.21.

    Therefore, $0.010 < P \text{ value} < 0.025$. Using a TI-84, $P \text{ value} \approx 0.0132$.

    (d) Since the P value is less than the level of significance $\alpha = 0.05$, we reject the null hypothesis.

    (e) At the 5% level of significance, there is sufficient evidence to conclude that Myers-Briggs preference and the profession are not independent.

11. **(a)** $\alpha = 0.01$

    H_0: Site type and pottery type are independent.

    H_1: Site type and pottery type are not independent.

    (b) $\chi^2 = \sum \dfrac{(O-E)^2}{E}$

    $$= \frac{(75-74.64)^2}{74.64} + \frac{(61-59.89)^2}{59.89} + \frac{(53-54.47)^2}{54.47} + \frac{(81-84.11)^2}{84.11} + \frac{(70-67.5)^2}{67.5}$$

    $$+ \frac{(62-61.39)^2}{61.39} + \frac{(92-89.25)^2}{89.25} + \frac{(68-71.61)^2}{71.61} + \frac{(66-65.14)^2}{65.14}$$

    $$= 0.5552$$

All expected frequencies are greater than 5. Use the chi-square distribution.

Since there are 3 rows and 3 columns, $d.f. = (3-1)(3-1) = 4$.

(c) In Table 7 in Appendix II, with $d.f. = 4$, $\chi^2 = 0.5552$ falls between entries 0.484 and 0.711.

Therefore, $0.950 < P$ value < 0.975. Using a TI-84, P value ≈ 0.9679.

(d) Since the P value is greater than the level of significance $\alpha = 0.01$, we do not reject the null hypothesis.

(e) At the 1% level of significance, there is insufficient evidence to conclude that site and pottery type are not independent.

13. **(a)** $\alpha = 0.05$

H_0: Age distribution and location are independent.

H_1: Age distribution and location are not independent.

(b) $\chi^2 = \sum \dfrac{(O-E)^2}{E}$

$$= \frac{(13-14.08)^2}{14.08} + \frac{(13-12.84)^2}{12.84} + \frac{(15-14.08)^2}{14.08} + \frac{(10-11.33)^2}{11.33} + \frac{(11-10.34)^2}{10.34}$$
$$+ \frac{(12-11.33)^2}{11.33} + \frac{(34-31.59)^2}{31.59} + \frac{(28-28.82)^2}{28.82} + \frac{(30-31.59)^2}{31.59}$$

$= 0.67$

All expected frequencies are greater than 5. Use the chi-square distribution.

Since there are 3 rows and 3 columns, $d.f. = (3-1)(3-1) = 4$.

(c) In Table 7 in Appendix II, with $d.f. = 4$, $\chi^2 = 0.67$ falls between entries 0.484 and 0.711.

Therefore, $0.950 < P$ value < 0.975. Using a TI-84, P value ≈ 0.9549.

(d) Since the P value is greater than the level of significance $\alpha = 0.05$, we do not reject the null hypothesis.

(e) At the 5% level of significance, there is insufficient evidence to conclude that age distribution and location are not independent.

15. **(a)** $\alpha = 0.05$

H_0: Ages of young adult and movie preference are independent.

H_1: Ages of young adult and movie preference are not independent.

(b) $\chi^2 = \sum \dfrac{(O-E)^2}{E}$

$$= \frac{(8-10.60)^2}{10.60} + \frac{(15-12.06)^2}{12.06} + \frac{(11-11.33)^2}{11.33} + \frac{(12-9.35)^2}{9.35} + \frac{(10-10.65)^2}{10.65}$$
$$+ \frac{(8-10.00)^2}{10.00} + \frac{(9-9.04)^2}{9.04} + \frac{(8-10.29)^2}{10.29} + \frac{(12-9.67)^2}{9.67}$$

$= 3.623$

All expected frequencies are greater than 5. Use the chi-square distribution.

Since there are 3 rows and 3 columns, $d.f. = (3-1)(3-1) = 4$.

(c) In Table 7 in Appendix II, with $d.f. = 4$, $\chi^2 = 3.623$ falls between entries 1.064 and 7.78.

Therefore, $0.100 < P$ value < 0.900. Using a TI-84, P value ≈ 0.4594.

(d) Since the P value is greater than the level of significance $\alpha = 0.05$, we do not reject the null hypothesis.

(e) At the 5% level of significance, there is insufficient evidence to conclude that age of young adult and movie preference are not independent.

17. **(a)** $\alpha = 0.05$
    H_0: Stone tool construction material and site are independent.
    H_1: Stone tool construction material and site are not independent.

(b) $\chi^2 = \sum \dfrac{(O-E)^2}{E}$

$$= \dfrac{(731-689.36)^2}{689.36} + \dfrac{(584-625.64)^2}{625.64} + \dfrac{(102-102.22)^2}{102.22} + \dfrac{(93-92.78)^2}{92.78}$$

$$+ \dfrac{(510-542.58)^2}{542.58} + \dfrac{(525-492.42)^2}{492.42} + \dfrac{(85-93.84)^2}{93.84} + \dfrac{(94-85.16)^2}{85.16}$$

$$= 11.15$$

All expected frequencies are greater than 5. Use the chi-square distribution.
Since there are 4 rows and 2 columns, $d.f. = (4-1)(2-1) = 3$.

(c) In Table 7 in Appendix II, with $d.f. = 3$, $\chi^2 = 11.15$ falls between entries 9.35 and 11.34.
Therefore, $0.010 < P$ value < 0.025. Using a TI-84, P value ≈ 0.0110.

(d) Since the P value is less than the level of significance $\alpha = 0.05$, we reject the null hypothesis.

(e) At the 5% level of significance, there is sufficient evidence to conclude that stone tool construction material and site are not independent.

19. **(i)**

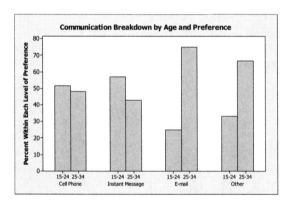

(ii) $\alpha = 0.05$
H_0: The proportion of age groups within each communication preference is the same.
H_1: The proportion of age groups within each communication preference is not the same.

$$\chi^2 = \dfrac{(O-E)^2}{E} = \dfrac{(48-44.5)^2}{44.5} + \dfrac{(45-44.5)^2}{44.5} + \dfrac{(40-35)^2}{35} + \dfrac{(30-35)^2}{35} +$$

$$\dfrac{(5-10)^2}{10} + \dfrac{(15-10)^2}{10} + \dfrac{(7-10.5)^2}{10.5} + \dfrac{(14-10.5)^2}{10.5} = 9.312$$

All expected frequencies are greater than 5. Use the chi-square distribution. Since there are 2 rows and 4 columns, $d.f. = (2-1)(4-1) = 3$. In Table 7 in Appendix II, with $d.f. = 3$, $\chi^2 = 9.312$ falls between entries 7.81 and 9.35. Therefore, $0.025 < P$ value < 0.050. Using a TI-84, P value ≈ 0.0254.

Since the P value is less than the level of significance $\alpha = 0.05$, we reject the null hypothesis.
At the 5% level of significance, there is sufficient evidence to conclude that the proportion of age groups within each communication preference is not the same.

Section 10.2

1. The degrees of freedom are the number of categories minus one.

3. The greater the differences between the observed frequencies and the expected frequencies, the larger the χ^2 value becomes. Larger χ^2 values lead to the conclusion that the differences are too large to be explained by chance alone.

5. **(a)** $\alpha = 0.05$
   H_0: The distributions are the same.
   H_1: The distributions are different.

   (b) $\chi^2 = \sum \dfrac{(O-E)^2}{E}$

   $= \dfrac{(47-32.76)^2}{32.76} + \dfrac{(75-61.88)^2}{61.88} + \dfrac{(288-305.31)^2}{305.31} + \dfrac{(45-55.06)^2}{55.06}$

   $= 11.79$

   All expected frequencies are greater than 5. Use the chi-square distribution.
   $d.f. = k - 1 = 4 - 1 = 3$.

   (c) In Table 7, with $d.f. = 3$, $\chi^2 = 11.79$ falls between entries 11.34 and 12.84.
   Therefore, $0.005 < P\text{-value} < 0.010$. Using a TI-84, $P\text{-value} \approx 0.0081$.

   (d) Since the P-value is less than the level of significance $\alpha = 0.05$, we reject the null hypothesis.

   (e) At the 5% level of significance, the evidence is sufficient to conclude that the Red Lake Village population does not fit the general Canadian population.

7. **(a)** $\alpha = 0.01$
   H_0: The distributions are the same.
   H_1: The distributions are different.

   (b) $\chi^2 = \sum \dfrac{(O-E)^2}{E}$

   $= \dfrac{(906-910.92)^2}{910.92} + \dfrac{(162-157.52)^2}{157.52} + \dfrac{(168-169.40)^2}{169.40} + \dfrac{(197-194.67)^2}{194.67} + \dfrac{(53-53.50)^2}{53.50}$

   $= 0.1984$

   All expected frequencies are greater than 5. Use the chi-square distribution.
   $d.f. = k - 1 = 5 - 1 = 4$.

   (c) In Table 7, with $d.f. = 4$, the entry 0.207 has 0.995 area in the right tail. Note that as χ^2 decrease, the area in the right tail increases so a $\chi^2 < 0.207$ means the corresponding P-value > 0.995. Using a TI-84, $P\text{-value} \approx 0.9953$.

   (d) Since the P-value is greater than the level of significance $\alpha = 0.01$, we do not reject the null hypothesis.

   (e) At the 1% level of significance, the evidence is insufficient to conclude that the regional distribution of raw materials does not fit the distribution at the current excavation site.

9. **(i)** Essay.

   (ii) **(a)** $\alpha = 0.01$

H_0: The distributions are the same.
H_1: The distributions are different.

(b) $\chi^2 = \sum \dfrac{(O-E)^2}{E}$

$$= \dfrac{(16-14.57)^2}{14.57} + \dfrac{(78-83.70)^2}{83.70} + \dfrac{(212-210.80)^2}{210.80} + \dfrac{(221-210.80)^2}{210.80}$$
$$+ \dfrac{(81-83.70)^2}{83.70} + \dfrac{(12-14.57)^2}{14.57}$$
$$= 1.5693$$

All expected frequencies are greater than 5. Use the chi-square distribution.
$d.f. = k - 1 = 6 - 1 = 5.$

(c) In Table 7, with $d.f. = 5$, $\chi^2 = 1.5693$ falls between entries 1.145 and 1.61.
Therefore, $0.900 < P\text{-value} < 0.950$. Using a TI-84, P-value ≈ 0.905.

(d) Since the P-value is greater than the level of significance $\alpha = 0.01$, we do not reject the null hypothesis.

(e) At the 1% level of significance, the evidence is insufficient to conclude that the average daily July temperature does not follow a normal distribution.

11. (a) $\alpha = 0.05$
H_0: The distributions are the same.
H_1: The distributions are different.

(b) $\chi^2 = \sum \dfrac{(O-E)^2}{E}$

$$= \dfrac{(120-150)^2}{150} + \dfrac{(85-75)^2}{75} + \dfrac{(220-200)^2}{200} + \dfrac{(75-75)^2}{75}$$
$$= 9.333$$

All expected frequencies are greater than 5. Use the chi-square distribution.
$d.f. = k - 1 = 4 - 1 = 3.$

(c) In Table 7, with $d.f. = 3$, $\chi^2 = 9.333$ falls between entries 7.81 and 9.35.
Therefore, $0.025 < P\text{-value} < 0.050$. Using a TI-84, P-value ≈ 0.0252.

(d) Since the P-value is less than the level of significance $\alpha = 0.05$, we reject the null hypothesis.

(e) At the 5% level of significance, the evidence is sufficient to conclude that current fish distribution is different than that of five years ago.

13. (a) $\alpha = 0.01$
H_0: The distributions are the same.
H_1: The distributions are different.

(b) $\chi^2 = \sum \dfrac{(O-E)^2}{E} = \dfrac{(127-121.50)^2}{121.50} + \dfrac{(40-36.45)^2}{36.45} + \dfrac{(480-461.70)^2}{461.70}$

$+ \dfrac{(502-498.15)^2}{498.15} + \dfrac{(56-72.90)^2}{72.90} + \dfrac{(10-24.30)^2}{24.30}$

$= 13.7$

All expected frequencies are greater than 5. Use the chi-square distribution.
$d.f. = k - 1 = 6 - 1 = 5.$

(c) In Table 7 in Appendix II, with $d.f. = 5$, $\chi^2 = 13.7$ falls between entries 12.83 and 15.09.
Therefore, $0.010 < P$ value < 0.025. Using a TI-84, P value ≈ 0.0176.

(d) Since the P value is greater than the level of significance $\alpha = 0.01$, we do not reject the null hypothesis.

(e) At the 1% level of significance, the evidence is insufficient to conclude that census distribution and the ethnic origin distribution of city residents are different.

15. **(a)** $\alpha = 0.01$
H_0: The distributions are the same.
H_1: The distributions are different.

(b) $\chi^2 = \sum \dfrac{(O-E)^2}{E}$

$= \dfrac{(83-82.775)^2}{82.775} + \dfrac{(49-48.4)^2}{48.4} + \dfrac{(32-34.375)^2}{34.375} + \dfrac{(22-26.675)^2}{26.675} + \dfrac{(25-21.725)^2}{21.725}$

$+ \dfrac{(18-18.425)^2}{18.425} + \dfrac{(13-15.95)^2}{15.95} + \dfrac{(17-14.025)^2}{14.025} + \dfrac{(16-12.65)^2}{12.65}$

$= 3.559$

All expected frequencies are greater than 5. Use the chi-square distribution.
$d.f. = k - 1 = 9 - 1 = 8.$

(c) In Table 7, with $d.f. = 8$, $\chi^2 = 3.559$ falls between entries 3.49 and 13.36.
Therefore, $0.100 < P$-value < 0.900. Using a TI-84, P-value ≈ 0.8946.

(d) Since the P-value interval is greater than the level of significance $\alpha = 0.01$, we do not reject the null hypothesis.

(e) At the 1% level of significance, the evidence is insufficient to conclude that the distribution of first-non-zero digits in the accounting file do not follow Benford's Law.

17. **(a)** $P(r) = \dfrac{e^{-\lambda}\lambda^r}{r!}$

$P(0) = \dfrac{e^{-1.72}(1.72)^0}{0!} \approx 0.179$

$$P(1) = \frac{e^{-1.72}(1.72)^1}{1!} \approx 0.308$$

$$P(2) = \frac{e^{-1.72}(1.72)^2}{2!} \approx 0.265$$

$$P(3) = \frac{e^{1.72}(1.72)^3}{3!} \approx 0.152$$

$$P(4 \text{ or more}) = 1 - P(3 \text{ or less})$$
$$= 1 - (0.179 + 0.308 + 0.265 + 0.152)$$
$$= 0.096$$

(b)

r	$E = 90P(r)$
0	$90(0.179) = 16.11$
1	$90(0.308) = 27.72$
2	$90(0.265) = 23.85$
3	$90(0.152) = 13.68$
4 or more	$90(0.096) = 8.64$

(c) $\chi^2 = \sum \dfrac{(O-E)^2}{E}$

$$= \frac{(22-16.11)^2}{16.11} + \frac{(21-27.72)^2}{27.72} + \frac{(15-23.85)^2}{23.85}$$
$$+ \frac{(17-13.68)^2}{13.68} + \frac{(15-8.64)^2}{8.64}$$
$$= 12.55$$

$d.f. = k - 1 = 5 - 1 = 4$

(d) $\alpha = 0.01$

H_0: The Poisson distribution fits.

H_1: The Poisson distribution does not fit.

In Table 7 in Appendix II, with $d.f. = 4$, $\chi^2 = 12.55$ falls between entries 11.14 and 13.28.
Therefore, $0.010 < P$ value < 0.025. Using a TI-84, P value ≈ 0.0137.
Since the P value is greater than the level of significance $\alpha = 0.01$, we do not reject the null hypothesis.
At the 1% level of significance, we cannot say that the Poisson distribution does not fit.

Section 10.3

1. Yes, it needs to be normal. No, the chi-square test of variance requires the x distribution to be exactly a normal distribution.

3. **(a)** $\alpha = 0.05$
    H_0: $\sigma^2 = 42.3$
    H_1: $\sigma^2 > 42.3$

    (b) $\chi^2 = \dfrac{(n-1)s^2}{\sigma^2} = \dfrac{(23-1)46.1}{42.3} = 23.98$; $d.f. = 23 - 1 = 22$

    Assume a normal population distribution.

   (c)   Since > is in H_1, a right-tailed test is used.

          Using Table 7 in Appendix II and $d.f. = 22$, $\chi^2 = 23.98$ falls between entries 14.04 and 30.81.

          Therefore, $0.100 < P$ value < 0.900. Using a TI-84, P value ≈ 0.3483.

   (d)   Since the P value is greater than the level of significance $\alpha = 0.05$, we do not reject the null hypothesis.

   (e)   At the 5% level of significance, there is insufficient evidence to conclude that the variance is greater in the new section.

   (f)   For $d.f. = 22$ and $\alpha = \dfrac{1-0.95}{2} = 0.025$, $\chi^2_U = 36.78$.

          For $d.f. = 22$ and $\alpha = \dfrac{1+0.95}{2} = 0.975$, $\chi^2_L = 10.98$.

          The 95% confidence interval for σ^2 is

$$\frac{(n-1)s^2}{\chi^2_U} < \sigma^2 < \frac{(n-1)s^2}{\chi^2_L}$$

$$\frac{(23-1)46.1}{36.78} < \sigma^2 < \frac{(23-1)46.1}{10.98}$$

$$27.57 < \sigma^2 < 92.37$$

5.   **(a)** $\alpha = 0.01$

          H_0: $\sigma^2 = 136.2$

          H_1: $\sigma^2 < 136.2$

   (b) $\chi^2 = \dfrac{(n-1)s^2}{\sigma^2} = \dfrac{(8-1)115.1}{136.2} = 5.92$; $d.f. = 8 - 1 = 7$

          Assume a normal population distribution.

   (c)   Since < is in H_1, a left-tailed test is used.

          Using Table 7 in Appendix II and $d.f. = 7$, $\chi^2 = 5.92$ falls between entries 2.83 and 12.02.

          Therefore, $0.900 <$ right-tail area < 0.100

          $1 - 0.900 < P$ value $< 1 - 0.100$

          $0.100 < P$ value < 0.900

          Using a TI-84, P value ≈ 0.4509.

   (d)   Since the P value is greater than the level of significance, we do not reject the null hypothesis.

   (e)   At the 1% level of significance, there is insufficient evidence to conclude that the variance for number of mountain climber deaths is less than 136.2

   (f)   For $d.f. = 7$ and $\alpha = \dfrac{1-0.90}{2} = 0.05$, $\chi^2_U = 14.07$.

          For $d.f. = 7$ and $\alpha = \dfrac{1+0.90}{2} = 0.95$, $\chi^2_L = 2.17$.

          The 90% confidence interval for σ^2 is

$$\frac{(n-1)s^2}{\chi^2_U} < \sigma^2 < \frac{(n-1)s^2}{\chi^2_L}$$

$$\frac{(8-1)115.1}{14.07} < \sigma^2 < \frac{(8-1)115.1}{2.17}$$

$$57.26 < \sigma^2 < 371.29$$

7. **(a)** $\alpha = 0.05$
    $H_0: \sigma^2 = 9$
    $H_1: \sigma^2 < 9$

    (b) $\chi^2 = \dfrac{(n-1)s^2}{\sigma^2} = \dfrac{(23-1)(1.9)^2}{3^2} = 8.82$; $d.f. = 23 - 1 = 22$

    Assume a normal population distribution.

    (c) Since $<$ is in H_1, a left-tailed test is used.

    Using Table 7 in Appendix II and $d.f. = 22$, $\chi^2 = 8.82$ falls between entries 8.64 and 9.54.

    Therefore, $0.995 <$ right-tail area < 0.990
    $1 - 0.995 < P$ value $< 1 - 0.990$
    $0.005 < P$ value < 0.010
    Using a TI-84, P value ≈ 0.0058.

    (d) Since the P value is less than the level of significance, we reject the null hypothesis.

    (e) At the 5% level of significance, there is sufficient evidence to conclude that the variance of protection times for the new typhoid shot is less than 9.

    (f) For $d.f. = 22$ and $\alpha = \dfrac{1-0.90}{2} = 0.05$, $\chi^2_U = 33.92$.

    For $d.f. = 22$ and $\alpha = \dfrac{1+0.90}{2} = 0.95$, $\chi^2_L = 12.34$.

    The 90% confidence interval for σ is

    $$\sqrt{\dfrac{(n-1)s^2}{\chi^2_U}} < \sigma < \sqrt{\dfrac{(n-1)s^2}{\chi^2_L}}$$

    $$\sqrt{\dfrac{(23-1)(1.9)^2}{33.92}} < \sigma < \sqrt{\dfrac{(23-1)(1.9)^2}{12.34}}$$

    $$1.53 < \sigma < 2.54$$

9. **(a)** $\alpha = 0.01$
    $H_0: \sigma^2 = 0.18$
    $H_1: \sigma^2 > 0.18$

    (b) $\chi^2 = \dfrac{(n-1)s^2}{\sigma^2} = \dfrac{(61-1)0.27}{0.18} = 90$; $d.f. = 61 - 1 = 60$

    Assume a normal population distribution.

    (c) Since $>$ is in H_1, a right-tailed test is used.

    Using Table 7 in Appendix II and $d.f. = 60$, $\chi^2 = 90$ falls between entries 88.38 and 91.95.

    Therefore, $0.005 < P$ value < 0.010. Using a TI-84, P value ≈ 0.0073.

    (d) Since the P value is less than the level of significance, we reject the null hypothesis.

    (e) At the 1% level of significance, there is sufficient evidence to conclude that the variance of measurements on the fan blades is higher than the specified amount. The inspector is justified in claiming the blades must be replaced.

(f) For $d.f. = 60$ and $\alpha = \dfrac{1-0.90}{2} = 0.05$, $\chi^2_U = 79.08$.

For $d.f. = 60$ and $\alpha = \dfrac{1+0.90}{2} = 0.95$, $\chi^2_L = 43.19$.

The 90% confidence interval for σ is

$$\sqrt{\frac{(n-1)s^2}{\chi^2_U}} < \sigma < \sqrt{\frac{(n-1)s^2}{\chi^2_L}}$$

$$\sqrt{\frac{(61-1)0.27}{79.08}} < \sigma < \sqrt{\frac{(61-1)0.27}{43.19}}$$

$$0.45 \text{ mm} < \sigma < 0.61 \text{ mm}$$

11. (i) (a) $\alpha = 0.05$
$H_0: \sigma^2 = 23$
$H_1: \sigma^2 \neq 23$

(b) $\chi^2 = \dfrac{(n-1)s^2}{\sigma^2} = \dfrac{(22-1)(14.3)}{23} = 13.06$; $d.f. = 22 - 1 = 21$

Assume a normal population distribution.

(c) The area to the left of $\chi^2 = 13.8$ is less than 50%, so we double the left-tail area to find the *P* value for the two-tailed test. Right-tail area is between 0.950 and 0.900. Subtracting each value from 1, we find the left-tail area is between 0.050 and 0.100. Doubling the left-tail area for a two-tailed test gives $0.100 < P$ value < 0.200.

(d) Since the *P* value is greater than the level of significance $\alpha = 0.05$, we do not reject the null hypothesis.

(e) At the 5% level of significance, there is insufficient evidence to conclude that the variance of battery life is different from 23.

(ii) For $d.f. = 21$ and $\alpha = \dfrac{1-0.90}{2} = 0.05$, $\chi^2_U = 32.67$.

For $d.f. = 21$ and $\alpha = \dfrac{1+0.90}{2} = 0.95$, $\chi^2_L = 11.59$.

The 90% confidence interval for σ^2 is

$$\frac{(n-1)s^2}{\chi^2_U} < \sigma^2 < \frac{(n-1)s^2}{\chi^2_L}$$

$$\frac{(22-1)(14.3)}{32.67} < \sigma^2 < \frac{(22-1)(14.3)}{11.59}$$

$$9.19 < \sigma^2 < 25.91$$

(iii) The 90% confidence interval for σ is

$$\sqrt{\frac{(n-1)s^2}{\chi^2_U}} < \sigma < \sqrt{\frac{(n-1)s^2}{\chi^2_L}}$$

$$\sqrt{9.19} < \sigma < \sqrt{25.91}$$

$$3.03 < \sigma < 5.09$$

Section 10.4

1. The populations should be independent.

3. F distributions are not symmetric. Values of the F distribution are always nonnegative.

5. **(a)** $\alpha = 0.01$. Population I is annual production from the first plot.
   H_0: $\sigma_1^2 = \sigma_2^2$
   H_1: $\sigma_1^2 > \sigma_2^2$

   (b) Since $s^2 \approx 0.332$ is larger than $s^2 \approx 0.089$, we designate Population I as the first plot.

   $$F = \frac{s_1^2}{s_2^2} = \frac{0.332}{0.089} = 3.73 \ ; \quad d.f._N = n_1 - 1 = 16 - 1 = 15, \text{ and } d.f._D = n_2 - 1 = 16 - 1 = 15$$

   The populations follow independent normal distributions. The samples are random samples from each population.

   (c) Since $>$ is in H_1, a right-tailed test is used.

   From Table 8 in Appendix II, $F = 3.73$ falls between entries 3.52 and 5.54.
   Therefore, $0.001 < P$ value < 0.010. Using a TI-84, P value ≈ 0.0075.

   (d) Since the P value is less than the level of significance $\alpha = 0.01$, we reject the null hypothesis.

   (e) At the 1% level of significance, there is sufficient evidence to show that the variance in annual wheat production from the first plot is greater than that of the second plot.

7. **(a)** $\alpha = 0.05$. Population I has data from France.
   H_0: $\sigma_1^2 = \sigma_2^2$
   H_1: $\sigma_1^2 \neq \sigma_2^2$

   (b) Since $s^2 \approx 2.044$ is larger than $s^2 \approx 1.038$, we designate Population I as France.

   $$F = \frac{s_1^2}{s_2^2} = \frac{2.044}{1.038} = 1.97 \ ; \quad d.f._N = n_1 - 1 = 21 - 1 = 20, \text{ and } d.f._D = n_2 - 1 = 18 - 1 = 17$$

   The populations follow independent normal distributions. The samples are random samples from each population.

   (c) Since $\neq$ is in H_1, a two-tailed test is used.

   From Table 8 in Appendix II, $F = 1.97$ falls between entries 1.86 and 2.23.
   Therefore, $0.050 <$ right-tail area < 0.100
   $0.100 < P$ value < 0.200
   Using a TI-84, P value ≈ 0.1631.

   (d) Since the P value is greater than the level of significance $\alpha = 0.05$, we do not reject the null hypothesis.

   (e) At the 5% level of significance, there is insufficient evidence to show that the variance in corporate productivity of large companies in France and those in Germany differ. Volatility of corporate productivity does not appear to differ.

9. **(a)** $\alpha = 0.05$. Population I has data from aggressive growth companies.
   H_0: $\sigma_1^2 = \sigma_2^2$
   H_1: $\sigma_1^2 > \sigma_2^2$

(b) Since $s^2 \approx 348.43$ is larger than $s^2 \approx 137.31$, we designate Population I as aggressive growth.

$$F = \frac{s_1^2}{s_2^2} = \frac{348.43}{137.31} = 2.54 \;; \quad d.f._N = n_1 - 1 = 21 - 1 = 20, \text{ and } d.f._D = n_2 - 1 = 21 - 1 = 20$$

The populations follow independent normal distributions. The samples are random samples from each population.

(c) Since > is in H_1, a right-tailed test is used.

From Table 8 in Appendix II, $F = 2.54$ falls between entries 2.46 and 2.94.
Therefore, $0.010 < P$ value < 0.025. Using a TI-84, P value ≈ 0.0216.

(d) Since the P value is less than the level of significance $\alpha = 0.05$, we reject the null hypothesis.

(e) At the 5% level of significance, there is sufficient evidence to show that the variance in percentage annual returns for funds holding aggressive-growth small stocks is larger than that for funds holding value stocks.

11. (a) $\alpha = 0.05$. Population I has data from the new system.

$$H_0: \sigma_1^2 = \sigma_2^2$$

$$H_1: \sigma_1^2 \neq \sigma_2^2$$

(b) Since $s^2 \approx 58.4$ is larger than $s^2 \approx 31.6$, we designate Population I as the new system.

$$F = \frac{s_1^2}{s_2^2} = \frac{58.4}{31.6} = 1.85 \;; \quad d.f._N = n_1 - 1 = 31 - 1 = 30, \text{ and } d.f._D = n_2 - 1 = 25 - 1 = 24$$

The populations follow independent normal distributions. The samples are random samples from each population.

(c) Since $\neq$ is in H_1, a two-tailed test is used.

From Table 8 in Appendix II, $F = 1.85$ falls between entries 1.67 and 1.94.
Therefore, $0.050 <$ right-tail area < 0.100
$0.100 < P$ value < 0.200
Using a TI-84, P value ≈ 0.1266.

(d) Since the P value is greater than the level of significance $\alpha = 0.05$, we do not reject the null hypothesis.

(e) At the 5% level of significance, there is insufficient evidence to show that the variance in gasoline consumption for the two injection systems is different.

Section 10.5

1. (a) $\alpha = 0.01$

$$H_0: \mu_1 = \mu_2 = \mu_3$$

H_1: Not all the means are equal.

(b)

Site I	Site II	Site III
$n = 7$	$n = 4$	$n = 6$
$\sum x_1 = 286$	$\sum x_2 = 164$	$\sum x_3 = 176$
$\sum x_1^2 = 15{,}312$	$\sum x_2^2 = 8354$	$\sum x_3^2 = 7450$
$SS_1 = 3626.857$	$SS_2 = 1630$	$SS_3 = 2287.33\overline{3}$

$$\sum x_{\text{TOT}} = 286 + 164 + 176 = 626$$

$$\sum x_{\text{TOT}}^2 = 15,312 + 8,354 + 7,450 = 31,116$$

$$N = 7 + 4 + 6 = 17$$

$$k = 3$$

$$SS_{\text{TOT}} = \sum x_{\text{TOT}}^2 - \frac{\left(\sum x_{\text{TOT}}\right)^2}{N} = 31,116 - \frac{(626)^2}{17} = 8,064.470$$

$$SS_{\text{BET}} = \sum_{\text{All groups}} \left(\frac{\left(\sum x_i\right)^2}{n_i} \right) - \frac{\left(\sum x_{\text{TOT}}\right)^2}{N}$$

$$= \frac{(286)^2}{7} + \frac{(164)^2}{4} + \frac{(176)^2}{6} - \frac{(626)^2}{17} = 520.280$$

$$SS_W = SS_1 + SS_2 + SS_3 = 3,626.857 + 1,630 + 2,287.333 = 7,544.190$$

Check that $SS_{\text{TOT}} = SS_{\text{BET}} + SS_W$: $8,064.470 = 520.280 + 7,544.190$

$$d.f._{\text{BET}} = k - 1 = 3 - 1 = 2$$

$$d.f._W = N - k = 17 - 3 = 14$$

$$d.f._{\text{TOT}} = N - 1 = 17 - 1 = 16$$

$$MS_{\text{BET}} = \frac{SS_{\text{BET}}}{d.f._{\text{BET}}} = \frac{520.280}{2} = 260.14$$

$$MS_W = \frac{SS_W}{d.f._W} = \frac{7,544.190}{14} = 538.87$$

$$F = \frac{MS_{\text{BET}}}{MS_W} = \frac{260.14}{538.87} = 0.48$$

For $d.f._N = 2$ and $d.f._D = 14$.

(c) From Table 8 in Appendix II, $F = 0.48$ falls below the entry 2.73.
Therefore, P value > 0.100. Using a TI-84, P value ≈ 0.6270.

(d) Since the P value is greater than the level of significance $\alpha = 0.01$, do not reject H_0.

(e) At the 1% level of significance, there is insufficient evidence to conclude that the means are not all equal.

(f) Summary of ANOVA results

Source of Variation	Sum of Squares	Degrees of Freedom	MS	F Ratio	P Value	Test Decision
Between groups	520.280	2	260.14	0.48	>0.100	Do not reject H_0
Within groups	7544.190	14	538.87			
Total	8064.470	16				

3. (a) $\alpha = 0.05$

$H_0: \mu_1 = \mu_2 = \mu_3 = \mu_4$

H_1: Not all the means are equal.

(b) See Section 10.5 or solutions to Problems 1 and 2 for examples of calculations from formulas.

$$F = \frac{MS_{\text{BET}}}{MS_W} = \frac{29.879}{35.325} \approx 0.846; \quad d.f._N = 3 \text{ and } d.f._D = 18$$

(c)   From Table 8 in Appendix II, $F = 0.846$ falls below entry 2.42.
Therefore, P value > 0.100. Using a TI-84, P value ≈ 0.4867.

(d)   Since the P value is greater than the level of significance $\alpha = 0.05$, we do not reject H_0.

(e)   At the 5% level of significance, there is insufficient evidence to conclude the means are not all equal.

(f)   Summary of ANOVA results

Source of Variation	Sum of Squares	Degrees of Freedom	MS	F Ratio	P Value	Test Decision
Between groups	89.637	3	29.879	0.846	>0.100	Do not reject H_0
Within groups	635.827	18	35.324			
Total	725.464	21				

5.   (a) $\alpha = 0.05$
$H_0: \mu_1 = \mu_2 = \mu_3$
H_1: Not all the means are equal.

(b)   See Section 10.5 or solutions to Problems 1 and 2 for examples of calculations from formulas.

$$F = \frac{MS_{BET}}{MS_W} = \frac{651.58}{130.19} \approx 5.005 \, ; \quad d.f._N = 2 \text{ and } d.f._D = 9$$

(c)   From Table 8 in Appendix II, $F = 5.005$ falls between entries 4.26 and 5.71.
Therefore, $0.025 < P$ value < 0.050. Using a TI-84, P value ≈ 0.0346.

(d)   Since the P value is less than the level of significance $\alpha = 0.05$, we reject H_0.

(e)   At the 5% level of significance, there is sufficient evidence to conclude that the means are not all equal.

(f)   Summary of ANOVA results

Source of Variation	Sum of Squares	Degrees of Freedom	MS	F Ratio	P Value	Test Decision
Between groups	1,303.167	2	651.58	5.005	$0.025 < P$ value < 0.050	Reject H_0
Within groups	1,171.750	9	130.19			
Total	2,474.917	11				

7.   (a) $\alpha = 0.01$
$H_0: \mu_1 = \mu_2 = \mu_3$
H_1: Not all the means are equal.

(b)   See Section 10.5 or solutions to Problems 1 and 2 for examples of calculations from formulas.

$$F = \frac{MS_{BET}}{MS_W} = \frac{1.021}{3.039} \approx 0.336 \, ; \quad d.f._N = 2 \text{ and } d.f._D = 11$$

(c)   From Table 8 in Appendix II, $F = 0.336$ falls below entry 2.86.
Therefore, P value > 0.100. Using a TI-84, P value ≈ 0.7217.

(d)   Since the P value is greater than the level of significance $\alpha = 0.01$, we do not reject H_0.

(e)   At the 1% level of significance, there is insufficient evidence to conclude that the means are not all equal.

(f)   Summary of ANOVA results

Source of Variation	Sum of Squares	Degrees of Freedom	MS	F Ratio	P Value	Test Decision
Between groups	2.042	2	1.021	0.336	> 0.100	Do not reject H_0
Within groups	33.428	11	3.039			
Total	35.470	13				

9.   **(a)** $\alpha = 0.05$

$H_0: \mu_1 = \mu_2 = \mu_3 = \mu_4$

H_1: Not all the means are equal.

(b)   See Section 10.5 or solutions to Problems 1 and 2 for examples of calculations from formulas.

$$F = \frac{MS_{\text{BET}}}{MS_W} = \frac{79.408}{17.223} \approx 4.611; \quad d.f._N = 3 \text{ and } d.f._D = 15$$

(c)   From Table 8 in Appendix II, $F = 4.611$ falls between entries 4.15 and 5.42. Therefore, $0.010 < P$ value < 0.025. Using a TI-84, P value ≈ 0.0177.

(d)   Since the P value is less than the level of significance $\alpha = 0.05$, we reject H_0.

(e)   At the 5% level of significance, there is sufficient evidence to conclude that the means are not all equal.

(f)   Summary of ANOVA results

Source of Variation	Sum of Squares	Degrees of	MS	F Ratio	P Value	Test Decision
Between groups	238.225	3	79.408	4.611	$0.010 < P$ value < 0.025	Reject H_0
Within groups	258.340	15	17.223			
Total	496.565	18				

Section 10.6

1.   There are two factors. One factor is *walking device*, with three levels, and the other factor is *task*, with two levels. The data table has six cells.

3.   Since the P value is less than 0.01, there is a significant difference in mean cadence according to the factor *walking device used*. The critical value is $F_{0.01} = 6.01$. Since the sample $F = 30.94$ is greater than $F_{0.01}$, F lies in the critical region, and we reject H_0.

5.   **(a)**   There are two factors. One factor is *income level*, with four levels, and the other factor is *media type*, with five levels.

(b)   For income,

H_0: There is no difference in population mean index based on income level.

H_1: At least two income levels have different population mean indices.

$$F_{\text{income}} = \frac{MS_{\text{income}}}{MS_{\text{error}}} = \frac{359}{130} \approx 2.77$$

P value ≈ 0.088

At the 5% level of significance, do not reject H_0.

The data do not indicate any differences in population mean index according to income level.

(c) For media,

   H_0: No difference in population mean index by media type.

   H_1: At least two media types have different population mean indices.

   $$F_{\text{media}} = \frac{MS_{\text{media}}}{MS_{\text{error}}} = \frac{4}{130} \approx 0.03$$

   P value ≈ 0.998

   At the 5% level of significance, do not reject H_0.

   The data do not indicate any differences in population mean index according to media type.

7.

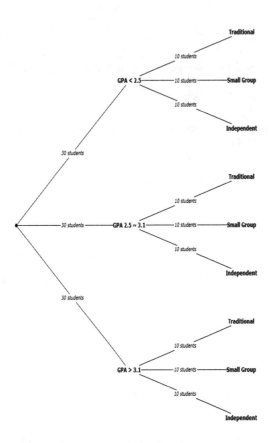

   Yes, the design fits the model for randomized block design.

Chapter Review Problems

1.   The chi-square and *F* distributions have only nonnegative values.

3.   Since we are interested in proportions, the test for homogeneity is appropriate.

5.   One-way ANOVA
   (i) $\alpha = 0.01$

      H_0: $\mu_1 = \mu_2 = \mu_3 = \mu_4$

      H_1: Not all the means are equal.

(ii) See Section 10.5 or solutions to Problems 1 and 2 for examples of calculations from formulas.

$$F = \frac{MS_{\text{BET}}}{MS_n} = \frac{2,050}{778} \approx 2.63 \, ; \quad d.f._{\cdot_N} = 3 \text{ and } d.f._{\cdot_D} = 16$$

(iii) From Table 8 in Appendix II, $F = 2.63$ falls between entries 2.46 and 3.24. Therefore, $0.050 < P$ value < 0.100. Using a TI-84, P value ≈ 0.0857.

(iv) Since the P value is greater than the level of significance $\alpha = 0.01$, we do not reject H_0.

(v) At the 1% level of significance, there is insufficient evidence to conclude that not all the packaging mean sales are equal.

(vi) Summary of ANOVA results

Source of Variation	Sum of Squares	Degrees of Freedom	MS	F Ratio	P Value	Test Decision
Between groups	6,150	3	2,050	2.63	$0.050 < P$ value < 0.100	Do not reject H_0
Within groups	12,455	16	778			
Total	18,605	19				

7. (a) Chi-square for testing σ^2

   (i) $\alpha = 0.01$
   H_0: $\sigma^2 = 1,040,400$
   H_1: $\sigma^2 > 1,040,400$

   (ii) $\chi^2 = \dfrac{(n-1)s^2}{\sigma^2} = \dfrac{(30-1)(1,353)^2}{(1,020)^2} = 51.03 \, ; \quad d.f. = 30 - 1 = 29$

   (iii) Since $>$ is in H_1, a right-tailed test is used.
   In Table 7 in Appendix II, with $d.f. = 29$, $\chi^2 = 51.03$ falls between entries 49.59 and 52.34. Therefore, $0.005 < P$ value < 0.010. Using a TI-84, P value ≈ 0.0070.

   (iv) Since the P value is less than the level of significance $\alpha = 0.01$, we reject the null hypothesis.

   (v) At the 1% level of significance, there is sufficient evidence to conclude that the variance is greater than claimed.

(b) For $d.f. = 29$ and $\alpha = \dfrac{1 - 0.95}{2} = 0.025$, $\chi_U^2 = 45.72$.

   For $d.f. = 29$ and $\alpha = \dfrac{1 + 0.95}{2} = 0.975$, $\chi_L^2 = 16.05$.

   The 95% confidence interval for σ^2 is

$$\frac{(n-1)s^2}{\chi_U^2} < \sigma^2 < \frac{(n-1)s^2}{\chi_L^2}$$

$$\frac{(30-1)(1,353)^2}{45.72} < \sigma^2 < \frac{(30-1)(1,353)^2}{16.05}$$

$$1,161,147.4 < \sigma^2 < 3,307,642.4 \text{ square foot-pounds}$$

9. Chi-square test of independence
   (i) $\alpha = 0.01$
   H_0: Student grade and teacher rating are independent.
   H_1: Student grade and teacher rating are not independent.

   (ii)
   $$\chi^2 = \sum \frac{(O-E)^2}{E}$$
   $$= \frac{(14-10.00)^2}{10.00} + \frac{(18-13.33)^2}{13.33} + \frac{(15-21.67)^2}{21.67} + \frac{(3-5.00)^2}{5.00} + \frac{(25-30.00)^2}{30.00}$$
   $$+ \frac{(35-40.00)^2}{40.00} + \frac{(75-65.00)^2}{65.00} + \frac{(15-15.00)^2}{15.00} + \frac{(21-20.00)^2}{20.00} + \frac{(27-26.67)^2}{26.67}$$
   $$+ \frac{(40-43.33)^2}{43.33} + \frac{(12-10.00)^2}{10.00}$$
   $$= 9.8$$

   Since there are 3 rows and 4 columns, $d.f. = (3-1)(4-1) = 6$.

   (iii) In Table 7 in Appendix II, with $d.f. = 6$, $\chi^2 = 9.8$ falls between entries 2.20 and 10.64.
   Therefore, $0.100 < P$ value < 0.900. Using a TI-84, P value ≈ 0.1333.
   (iv) Since the P value is greater than the level of significance $\alpha = 0.01$, we do not reject the null hypothesis.
   (v) At the 1% level of significance, there is insufficient evidence to claim that student grade and teacher rating are not independent.

11. Chi-square goodness of fit
    (i) $\alpha = 0.01$
    H_0: The distributions are the same.
    H_1: The distributions are different.

    (ii)
    $$\chi^2 = \sum \frac{(O-E)^2}{E} = \frac{(26-42.0)^2}{42.0} + \frac{(27-31.5)^2}{31.5} + \frac{(69-63.0)^2}{63.0}$$
    $$+ \frac{(68-52.5)^2}{52.5} + \frac{(20-21.0)^2}{21.0} = 11.93$$

    $$d.f. = k - 1 = 5 - 1 = 4$$

    (iii) In Table 7 in Appendix II, with 4 $d.f.$, $\chi^2 = 11.93$ falls between entries 11.14 and 13.28.
    Therefore, $0.010 < P$ value < 0.025. Using a TI-84, P value ≈ 0.0179.
    (iv) Since the P value is greater than the level of significance $\alpha = 0.01$, we do not reject H_0.
    (v) At the 1% level of significance, there is insufficient evidence to claim that the age distribution of the population of Blue Valley has changed.

13. *F* test for the equality of two variances.
    (i) $\alpha = 0.05$
    H_0: $\sigma_1^2 = \sigma_2^2$
    H_1: $\sigma_1^2 > \sigma_2^2$
    (ii) Since $s^2 = 135.24$ is larger than $s^2 = 51.87$, we designate Population I as the new process.
    $$F = \frac{s_1^2}{s_2^2} = \frac{135.24}{51.87} = 2.61; \quad d.f._N = n_1 - 1 = 16 - 1 = 15, \text{ and } d.f._D = n_2 - 1 = 18 - 1 = 17$$

(iii) Since $>$ is in H_1, a right-tailed test is used.

From Table 8 in Appendix II, $F = 2.61$ falls between entries 2.31 and 2.72.
Therefore, $0.025 < P$ value < 0.050. Using a TI-84, P value ≈ 0.0302.

(iv) Since the P value is less than the level of significance $\alpha = 0.05$, we reject the null hypothesis.

(v) At the 5% level of significance, there is sufficient evidence to show that the variance for the lifetimes of bulbs manufactured using the new process is larger than that for bulbs made by the old process.

Chapter 11: Nonparametric Statistics

Section 11.1

1. Dependent, matched pairs are required.

3.

Region	Modern %	Historic %	Sign of Difference
1	4.0	3.3	+
2	2.3	1.9	+
3	7.8	7.0	+
4	2.8	5.5	−
5	0.7	3.3	−
6	5.1	6.0	−
7	2.9	3.2	−
8	4.2	8.2	−
9	4.9	6.4	−
10	5.8	7.2	−
11	6.8	6.1	+
12	3.6	1.5	+
13	3.2	1.0	+
14	0.8	2.1	−
15	7.3	5.1	+

(a) $\alpha = 0.05$

H_0: Distributions are the same.

H_1: Distributions are different.

(b) $x = \dfrac{\text{number of plus signs}}{\text{total number of signs}} = \dfrac{7}{15} \approx 0.4667$

$z = \dfrac{x - 0.5}{\sqrt{\frac{0.25}{n}}} = \dfrac{0.4667 - 0.5}{\sqrt{\frac{0.25}{15}}} \approx -0.26$

Use the standard normal distribution.

(c) By Table 5 in Appendix II, $P(z < -0.26) = 0.3974$. For a two-tailed test, P value $= 2(0.3974) = 0.7948$.

(d) Since the P value is greater than the level of significance $\alpha = 0.05$, do not reject H_0.

(e) At the 5% level of significance, the data are not significant. The evidence is insufficient to conclude that the economies are different.

5.

Student	After	Before	Sign of Difference
1	107	111	−
2	115	110	+
3	120	93	+
4	78	75	+
5	83	88	−
6	56	56	N.D.
7	71	75	−
8	89	73	+
9	77	83	−
10	44	40	+
11	119	115	+
12	130	101	+
13	91	110	−
14	99	90	+
15	96	98	−
16	83	76	+
17	100	100	N.D.
18	118	109	+

(a) $\alpha = 0.05$

H_0: Distributions are the same.

H_1: Distributions are different.

(b) $x = \dfrac{\text{number of plus signs}}{\text{total number of signs}} = \dfrac{10}{16} = 0.625$

$z = \dfrac{x - 0.5}{\sqrt{\frac{0.25}{n}}} = \dfrac{0.625 - 0.5}{\sqrt{\frac{0.25}{16}}} = 1.00$

Use the standard normal distribution.

(c) By Table 5 in Appendix II, $P(z > 1.00) = 0.1587$. For a two-tailed test, P value $= 2(0.1587) = 0.3174$.

(d) Since the P value is greater than the level of significance $\alpha = 0.05$, do not reject H_0.

(e) At the 5% level of significance, the data are not significant. The evidence is insufficient to conclude that the lectures have any effect on student awareness of current events.

7.

Twin Pair	A	B	Sign of Difference
1	177	86	+
2	150	135	+
3	112	115	−
4	95	110	−
5	120	116	+
6	117	84	+
7	86	93	−
8	111	77	+
9	110	96	+
10	142	130	+
11	125	147	−
12	89	101	−

(a) $\alpha = 0.05$

H_0: Distributions are the same.

H_1: Distributions are different.

(b) $x = \dfrac{\text{number of plus signs}}{\text{total number of signs}} = \dfrac{7}{12} \approx 0.5833$

$z = \dfrac{x - 0.5}{\sqrt{\frac{0.25}{n}}} = \dfrac{0.5833 - 0.5}{\sqrt{\frac{0.25}{12}}} \approx 0.58$

Use the standard normal distribution.

(c) By Table 5 in Appendix II, $P(z > 0.58) = 0.2810$. For a two-tailed test, P value $= 2(0.2810) = 0.5620$.

(d) Since the P value is greater than the level of significance $\alpha = 0.05$, do not reject H_0.

(e) At the 5% level of significance, the data are not significant. The evidence is insufficient to conclude that the schools are not equally effective.

9.

Subject	After	Before	Sign of Difference
1	28	28	N.D.
2	15	35	−
3	2	14	−
4	20	20	N.D.
5	31	25	+
6	19	40	−
7	6	18	−
8	17	15	+
9	1	21	−
10	5	19	−
11	12	32	−
12	20	42	−
13	30	26	+
14	19	37	−
15	0	19	−
16	16	38	−
17	4	23	−
18	19	24	−

(a) $\alpha = 0.01$

H_0: Distributions are the same.

H_1: Distribution after hypnosis is lower.

(b) $x = \dfrac{\text{number of plus signs}}{\text{total number of signs}} = \dfrac{3}{16} = 0.1875$

$z = \dfrac{x - 0.5}{\sqrt{\frac{0.25}{n}}} = \dfrac{0.1875 - 0.5}{\sqrt{\frac{0.25}{16}}} = -2.5$

Use the standard normal distribution.

(c) By Table 5 in Appendix II, the area in the left tail is $P(z < -2.5) = 0.0062$. P value $= 0.0062$.

(d) Since the P value is less than the level of significance $\alpha = 0.01$, reject H_0.

(e) At the 1% level of significance, the data are significant. The evidence is sufficient to conclude that the number of cigarettes smoked per day was less after hypnosis.

11.

Region	Male	Female	Sign of Difference
1	7.3	7.5	−
2	7.5	6.4	+
3	7.7	6.0	+
4	21.8	20.0	+
5	4.2	2.6	+
6	12.2	5.2	+
7	3.5	3.1	+
8	4.2	4.9	−
9	8.0	12.1	−
10	9.7	10.8	−
11	14.1	15.6	−
12	3.6	6.3	−
13	3.6	4.0	−
14	4.0	3.9	+
15	5.2	9.8	−
16	6.9	9.8	−
17	15.6	12.0	+
18	6.3	3.3	+
19	8.0	7.1	+
20	6.5	8.2	−

(a) $\alpha = 0.01$

H_0: Distributions are the same.

H_1: Distributions are different.

(b) $x = \dfrac{\text{number of plus signs}}{\text{total number of signs}} = \dfrac{10}{20} = 0.500$

$z = \dfrac{x - 0.5}{\sqrt{\frac{0.25}{n}}} = \dfrac{0.500 - 0.5}{\sqrt{\frac{0.25}{20}}} = 0$

Use the standard normal distribution.

(c) By Table 5 in Appendix II, $P(z > 0) = 0.5000$. For a two-tailed test, P value $= 2(0.5000) = 1.000$.

(d) Since the P value is greater than the level of significance $\alpha = 0.01$, do not reject H_0.

(e) At the 1% level of significance, the data are not significant. The evidence is insufficient to conclude that distribution of dropout rates is different for males and females.

Section 11.2

1. Independent samples are required.

3.
Yield	Method	Rank
1.10	B	1
1.15	A	2
1.25	A	3
1.34	B	4
1.41	A	5
1.53	B	6
1.61	A	7
1.75	B	8
1.78	A	9
1.80	B	10
1.83	A	11
1.88	B	12
1.89	B	13
1.95	A	14
1.96	B	15
1.99	A	16
2.01	A	17
2.11	B	18
2.12	A	19
2.15	B	20
2.17	B	21
2.21	B	22
2.34	A	23

(a) $\alpha = 0.05$

H_0: Distributions are the same.

H_1: Distributions are different.

(b) Since $n_1 = 11$ and $n_2 = 12$, R is the sum of the ranks of group A.

$$R_A = 2+3+5+7+9+11+14+16+17+19+23 = 126$$

$$\mu_R = \frac{n_1(n_1+n_2+1)}{2} = \frac{11(11+12+1)}{2} = 132$$

$$\sigma_R = \sqrt{\frac{n_1 n_2(n_1+n_2+1)}{12}} = \sqrt{\frac{11(12)(11+12+1)}{12}} \approx 16.25$$

$$z = \frac{R - \mu_R}{\sigma_R} = \frac{126 - 132}{16.25} \approx -0.37$$

Use the standard normal distribution because n_1 and n_2 are each > 10.

(c) By Table 5 in Appendix II, $P(z < -0.37) = 0.3557$. For a two-tail test, P value $= 2(0.3557) = 0.7114$.

(d) Since the P value is greater than the level of significance, do not reject H_0.

(e) At the 5% level of significance, the evidence is insufficient to conclude that the yield distributions are different between organic and conventional farming methods.

5.

Sessions	Group	Rank
19	A	1
20	B	2
24	B	3
25	A	4
26	B	5
27	A	6
28	A	7
31	A	8
33	B	9
34	B	10
35	A	11
36	A	12
37	A	13
38	A	14
39	B	15
40	A	16
41	A	17
42	B	18
43	A	19
44	B	20
45	B	21
46	B	22
48	B	23

(a) $\alpha = 0.05$

H_0: Distributions are the same.

H_1: Distributions are different.

(b)   Since $n_1 = 12$ and $n_2 = 11$, R is the sum of the ranks of group B.

$$R_B = 2+3+5+9+10+15+18+20+21+22+23 = 148$$

$$\mu_R = \frac{n_1(n_1+n_2+1)}{2} = \frac{12(12+11+1)}{2} = 144$$

$$\sigma_R = \sqrt{\frac{n_1 n_2 (n_1+n_2+1)}{12}} = \sqrt{\frac{12(11)(12+11+1)}{12}} \approx 16.25$$

$$z = \frac{R-\mu_R}{\sigma_R} = \frac{148-144}{16.25} \approx 0.25$$

Use the standard normal distribution because n_1 and n_2 are each > 10.

(c)   By Table 5 in Appendix II, $P(z > 0.25) = 0.3974$. For a two-tail test, P value $= 2(0.3974) = 0.7948$.

(d)   Since the P value is greater than the level of significance $\alpha = 0.05$, do not reject H_0.

(e)   At the 5% level of significance, the evidence is insufficient to conclude that the distributions of training sessions are different.

7.

Minutes	Group	Rank
7	A	1.0
8	A	2.0
9	B	3.0
10	A	4.0
11	A	5.5
11	B	5.5
12	A	7.0
13	A	8.0
14	B	9.0
15	A	10.0
16	A	11.5
16	B	11.5
17	A	13.0
18	A	14.0
19	B	15.0
22	A	16.0
24	B	17.0
27	B	18.0
28	B	19.0
29	B	20.0
30	B	21.0
31	B	22.0
33	B	23.0

(a) $\alpha = 0.05$

H_0: Distributions are the same.

H_1: Distributions are different.

(b) Since $n_1 = 11$ and $n_2 = 12$, R is the sum of the ranks of group A.

$$R_A = 1+2+4+5.5+7+8+10+11.5+13+14+16 = 92$$

$$\mu_R = \frac{n_1(n_1+n_2+1)}{2} = \frac{11(11+12+1)}{2} = 132$$

$$\sigma_R = \sqrt{\frac{n_1 n_2 (n_1+n_2+1)}{12}} = \sqrt{\frac{11(12)(11+12+1)}{12}} \approx 16.25$$

$$z = \frac{R-\mu_R}{\sigma_R} = \frac{92-132}{16.25} \approx -2.46$$

Use the standard normal distribution because n_1 and n_2 are each > 10.

(c) By Table 5 in Appendix II, $P(z < -2.46) = 0.0069$. For a two-tail test, P value $= 2(0.0069) = 0.0138$.

(d) Since the P value is less than the level of significance $\alpha = 0.05$, reject H_0.

(e) At the 5% level of significance, the evidence is sufficient to conclude that the completion time distributions for the two settings are different.

9.

Percent	Group	Rank
29.1	B	1
33.7	B	2
35.7	B	3
36.9	B	4
38.2	B	5
40.1	B	6
44.2	B	7
44.3	A	8
45.2	B	9
46.6	B	10
47.1	A	11
49.1	A	12
50.0	A	13
53.3	B	14
55.1	A	15
57.3	A	16
58.2	A	17
59.6	B	18
59.9	A	19
60.0	A	20
60.2	B	21
63.3	A	22
68.7	A	23

(a) $\alpha = 0.01$

H_0: Distributions are the same.

H_1: Distributions are different.

(b)   Since $n_1 = 11$ and $n_2 = 12$, R is the sum of the ranks of group A.

$$R_A = 8 + 11 + 12 + 13 + 15 + 16 + 17 + 19 + 20 + 22 + 23 = 176$$

$$\mu_R = \frac{n_1(n_1 + n_2 + 1)}{2} = \frac{11(11 + 12 + 1)}{2} = 132$$

$$\sigma_R = \sqrt{\frac{n_1 n_2 (n_1 + n_2 + 1)}{12}} = \sqrt{\frac{11(12)(11 + 12 + 1)}{12}} \approx 16.25$$

$$z = \frac{R - \mu_R}{\sigma_R} = \frac{176 - 132}{16.25} \approx 2.71$$

Use the standard normal distribution because n_1 and n_2 are each > 10.

(c)   By Table 5 in Appendix II, $P(z > 2.71) = 0.0034$. For a two-tail test, P value = $2(0.0034) = 0.0068$.

(d)   Since the P value is less than the level of significance $\alpha = 0.01$, reject H_0.

(e)   At the 1% level of significance, the evidence is sufficient to conclude that the distributions showing percentage of exercisers differ by education level.

11.

Scores	Group	Rank
61	A	1.0
62	A	2.0
63	B	3.0
65	A	4.5
65	B	4.5
69	A	6.0
70	B	7.0
72	B	8.0
75	B	9.0
77	A	10.0
78	A	11.0
79	A	12.5
79	B	12.5
80	B	14.0
81	B	15.0
82	B	16.0
83	A	17.0
85	A	18.0
87	A	19.0
90	B	20.0
92	A	21.0
93	A	22.0
94	B	23.0
95	A	24.0

(a) $\alpha = 0.01$

   H_0: Distributions are the same.

   H_1: Distributions are different.

(b)   Since $n_1 = 12$ and $n_2 = 12$, R is the sum of the ranks of either group.

$$R_A = 1 + 4.5 + 6 + 10 + 11 + 12.5 + 17 + 18 + 19 + 21 + 22 + 24 = 166$$

$$\mu_R = \frac{n_1(n_1 + n_2 + 1)}{2} = \frac{12(12 + 12 + 1)}{2} = 150$$

$$\sigma_R = \sqrt{\frac{n_1 n_2 (n_1 + n_2 + 1)}{12}} = \sqrt{\frac{12(12)(12 + 12 + 1)}{12}} \approx 17.32$$

$$z = \frac{R - \mu_R}{\sigma_R} = \frac{166 - 150}{17.32} \approx 0.92$$

Use the standard normal distribution because n_1 and n_2 are each > 10.

(c)   By Table 5 in Appendix II, $P(z > 0.92) = 0.1788$. For a two-tail test, P value = $2(0.1788) = 0.3576$.

(d)   Since the P value is greater than the level of significance $\alpha = 0.01$, do not reject H_0.

(e)   At the 1% level of significance, the evidence is insufficient to conclude that the distributions of test scores differ according to instruction method.

Section 11.3

1. Monotone increasing

3.

Person	Class Rank x	Sales Rank y	$d = x - y$	d^2
1	6	4	2	4
2	8	9	−1	1
3	11	10	1	1
4	2	1	1	1
5	5	6	−1	1
6	7	7	0	0
7	3	8	−5	25
8	9	11	−2	4
9	1	3	−2	4
10	10	5	5	25
11	4	2	2	4
Sum	66	66	0	70

(a) $\alpha = 0.05$

$H_0: \rho_s = 0$ (There is no monotone relationship between x and y.)

$H_1: \rho_s \neq 0$ (There is a monotonic relationship between x and y.)

(b) $r_s = 1 - \dfrac{6 \sum d^2}{n(n^2 - 1)} = 1 - \dfrac{6(70)}{11(11^2 - 1)} \approx 0.682$

(c) From Table 9 in Appendix II, 0.682 falls between entries 0.619 and 0.764 in the $n = 11$ row. Use two-tailed areas to find that $0.01 < P$ value < 0.05.

(d) Since the P value is less than the level of significance $\alpha = 0.05$, reject H_0.

(e) At the 5% level of significance, we conclude that there is a monotonic relationship (either increasing or decreasing) between the rank in training class and the rank in sales.

5.

Rat Colony	Population Density Rank x	Violence Rank y	$d = x - y$	d^2
1	3	1	2	4
2	5	3	2	4
3	6	5	1	1
4	1	2	−1	1
5	8	8	0	0
6	7	6	1	1
7	4	4	0	0
8	2	7	−5	25
Sum	36	36	0	36

(a) $\alpha = 0.05$

$H_0: \rho_s = 0$ (There is no monotonic relationship.)

$H_1: \rho_s > 0$ (The relationship between x and y is monotone increasing.)

(b) $r_s = 1 - \dfrac{6 \sum d^2}{n(n^2 - 1)} = 1 - \dfrac{6(36)}{8(64 - 1)} = 0.571$

(c) From Table 9 in Appendix II, 0.571 falls to the left of the entry 0.620 in the $n = 8$ row. Use one-tail areas to find that the P value > 0.05.

(d) Since the P value is greater than the level of significance $\alpha = 0.05$, do not reject H_0.

(e) At the 5% level of significance, there is insufficient evidence to indicate a monotonic-increasing relationship between crowding and violence.

7. **(i)**

Soldier	Humor Test Score x	Humor Test Rank x	Aggressiveness Test Score	Aggressiveness Rank y	$d = x - y$	d^2
1	60	5	78	1	4	16
2	85	3	42	7	−4	16
3	78	4	68	3	1	1
4	90	2	53	5	−3	9
5	93	1	62	4	−3	9
6	45	7	50	6	1	9
7	51	6	76	2	4	1
Sum	—	28	—	28	0	68

(ii) **(a)** $\alpha = 0.05$

$H_0: \rho_s = 0$ (There is no monotonic relationship.)

$H_1: \rho_s < 0$ (There is a monotone-decreasing relationship between x and y.)

(Here, soldiers with a greater sense of humor have lower aggression scores.)

(b) $r_s = 1 - \dfrac{6 \sum d^2}{n(n^2 - 1)} = 1 - \dfrac{6(68)}{7(48)} \approx -0.214$

(c) From Table 9 in Appendix II, 0.214 falls to the left of entry 0.715 in the $n = 7$ row. Use the one-tailed areas to find that the P value > 0.05.

(d) Since the P value is greater than the level of significance $\alpha = 0.05$, do not reject H_0.

(e) At the 5% level of significance, the evidence is insufficient to conclude that there is a monotonic decreasing relationship between ranks of humor and ranks of aggressiveness.

9. **(i)**

Area	Police Rank x	Fire Fighter Rank y	$d = x - y$	d^2
1	12	12.0	0.0	0.00
2	7	8.0	−1.0	1.00
3	10	10.0	0.0	0.00
4	5	4.0	1.0	1.00
5	4	5.5	−1.5	2.25
6	2	2.0	0.0	0.00
7	6	5.5	0.5	0.25
8	9	11.0	−2.0	4.00
9	3	3.0	0.0	0.00
10	13	13.0	0.0	0.00
11	11	7.0	4.0	16.00
12	8	9.0	−1.0	1.00
13	1	1.0	0.0	0.00
Sum	91	91	0	25.5

(ii) **(a)** $\alpha = 0.05$

$H_0: \rho_s = 0$ (no monotone relationship)

$H_1: \rho_s \neq 0$ (monotone relationship)

(b) $r_s = 1 - \dfrac{6 \sum d^2}{n(n^2 - 1)} = 1 - \dfrac{6(25.5)}{13(13^2 - 1)} \approx 0.930$

(c)   From Table 9 in Appendix II, 0.930 falls to the right of the entry 0.797 in the $n = 13$ row. Use the two-tailed areas to find that the P value < 0.002.

(d)   Since the P value is less than the level of significance $\alpha = 0.05$, reject H_0.

(e)   At the 5% level of significance, we conclude that there is a monotonic relationship between number of fire fighters and number of police.

11.   (i)

City	Insurance Sales Rank x	Per Capita Income	Income Rank y	$d = x - y$	d^2
1	6	17	5	1	1
2	7	18	4	3	9
3	1	19	2.5	−1.5	2.25
4	8	11	8	0	0
5	3	16	6	−3	9
6	2	20	1	1	1
7	5	15	7	−2	4
8	4	19	2.5	1.5	2.25
Sum	36	—	36	0	28.5

(ii)   (a) $\alpha = 0.01$

$H_0: \rho_s = 0$ (There is no monotonic relationship.)

$H_1: \rho_s \neq 0$ (There is a monotone relationship between x and y.)

(b) $r_s = 1 - \dfrac{6 \sum d^2}{n(n^2 - 1)} = 1 - \dfrac{6(28.5)}{8(64 - 1)} \approx 0.661$

(c)   From Table 9 in Appendix II, 0.661 falls between entries 0.620 and 0.715 in the $n = 8$ row. Use two-tailed areas to find that $0.05 < P$ value < 0.10.

(d)   Since the P value is greater than the level of significance $\alpha = 0.01$, do not reject H_0.

(e)   At the 1% level of significance, we conclude that there is insufficient evidence to reject the null hypothesis of no monotonic relationship between rank of insurance sales and rank of per capita income.

Section 11.4

1.   Exactly two

3.   **(a)** $\alpha = 0.05$

H_0: The symbols are randomly mixed in the sequence.

H_1: The symbols are not randomly mixed in the sequence.

(b) $RRR|DD|RR|DDDDD|R|D|RR|D|RRR|DD|R$; $R = 11$

(c)   Letting R be the first symbol, $n_1 = 12$ and $n_2 = 11$. From Table 10 in Appendix II, $c_1 = 7$ and $c_2 = 18$.

(d)

$R \leq 7$	$8 \leq R \leq 17$	$R \geq 18$
Reject H_0	Fail to reject H_0	Reject H_0

Since $R = 11$, do not reject H_0.

(e)   At the 5% level of significance, the evidence is insufficient to conclude that the sequence of presidential party affiliation is not random.

5. **(a)** $\alpha = 0.05$
    H_0: The symbols are randomly mixed in the sequence.
    H_1: The symbols are not randomly mixed in the sequence.

    (b) $SSS|N|S|N|SSSS|NN|S|N|SSS|NN|SSSS$; $R = 11$

    (c)   Letting S be the first symbol, $n_1 = 16$ and $n_2 = 7$. From Table 10 in Appendix II, $c_1 = 6$ and $c_2 = 16$.

    (d)

$R \le 6$	$7 \le R \le 15$	$R \ge 16$
Reject H_0	Fail to reject H_0	Reject H_0

    Since $R = 11$, we do not reject H_0.

    (e)   At the 5% level of significance, the evidence is insufficient to conclude that the sequence of days for seeding and not seeding is not random.

7. **(i)**   Median = 11.7.
    Using A for above the median and B for below, the original sequence translates to $BBBAAAAABBBA$.

    (ii) **(a)** $\alpha = 0.05$
    H_0: The numbers are randomly mixed about the median.
    H_1: The numbers are not randomly mixed about the median.

    (b) $BBB|AAAAA|BBB|A$; $R = 4$

    (c)   Letting B be the first symbol, $n_1 = 6$ and $n_2 = 6$. From Table 10 in Appendix II, $c_1 = 3$ and $c_2 = 11$.

    (d)

$R \le 3$	$4 \le R \le 10$	$R \ge 11$
Reject H_0	Fail to reject H_0	Reject H_0

    Since $R = 4$, do not reject H_0.

    (e)   At the 5% level of significance, the evidence is insufficient to conclude that the sequence of returns is not random about the median.

9. **(i)**   Median = 21.6.
    Using A for above the sequence and B for below, the original sequence translates to $BAAAAAABBBBB$.

    (ii) **(a)** $\alpha = 0.05$
    H_0: The numbers are randomly mixed about the median.
    H_1: The numbers are not randomly mixed about the median.

    (b) $B|AAAAAA|BBBBB$; $R = 3$

    (c)   Letting B be the first symbol, $n_1 = 6$ and $n_2 = 6$. From Table 10 in Appendix II, $c_1 = 3$ and $c_2 = 11$.

    (d)

$R \le 3$	$4 \le R \le 10$	$R \ge 11$
Reject H_0	Fail to reject H_0	Reject H_0

    Since $R = 3$, reject H_0.

(e) At the 5% level of significance, we can conclude that the sequence of percentage of sand in the soil at successive depths is not random about the median.

11. (a) H_0: The symbols are randomly mixed in the sequence.

H_1: The symbols are not randomly mixed in the sequence.

(b) $n_1 = 21;\ n_2 = 17;\ R = 18$

(c) $\mu_R = \dfrac{2n_1 n_2}{n_1 + n_2} + 1 = \dfrac{2(21)(17)}{21 + 17} + 1 \approx 19.80$

$\sigma_R = \sqrt{\dfrac{(2n_1 n_2)(2n_1 n_2 - n_1 - n_2)}{(n_1 + n_2)^2 (n_1 + n_2 - 1)}} = \sqrt{\dfrac{2(21)(17)[2(21)(17) - 21 - 17]}{(21 + 17)^2 (21 + 17 - 1)}} \approx 3.01$

$z = \dfrac{R - \mu_R}{\sigma_R} = \dfrac{18 - 19.80}{3.01} \approx -0.60$

(d) Since $-1.96 < -0.60 < 1.96$, do not reject H_0; $P(z < -0.60) = 0.2743$; P value $\approx 2(0.2743) = 0.5486$; at the 5% level of significance, the P value also tells us not to reject H_0.

(e) At the 5% level of significance, the evidence is insufficient to reject the null hypothesis of a random sequence of Democrat and Republican presidential terms.

Chapter Review Problems

1. No assumptions about the population distribution are required.

3. (a) Rank-sum test

Index	Group	Rank
1.1	1	1
1.5	2	2
1.6	1	3
1.8	1	4
1.9	2	5
2.2	2	6
2.4	2	7
2.5	1	8
2.8	2	9
2.9	1	10
3.1	2	11
3.2	1	12
3.3	2	13
3.5	2	14
3.6	2	15
3.7	1	16
3.8	1	17
3.9	2	18
4.0	2	19
4.1	1	20
4.2	1	21
4.4	1	22
4.6	2	23

(b) $\alpha = 0.05$

H_0: Distributions are the same.

H_1: Distributions are different.

(c) Since $n_1 = 11$ and $n_2 = 12$, R is the sum of the ranks of group 1.

$$R_1 = 1 + 3 + 4 + 8 + 10 + 12 + 16 + 17 + 20 + 21 + 22 = 134$$

$$\mu_R = \frac{n_1(n_1 + n_2 + 1)}{2} = \frac{11(11 + 12 + 1)}{2} = 132$$

$$\sigma_R = \sqrt{\frac{n_1 n_2 (n_1 + n_2 + 1)}{12}} = \sqrt{\frac{11(12)(11 + 12 + 1)}{12}} \approx 16.25$$

$$z = \frac{R - \mu_R}{\sigma_R} = \frac{134 - 132}{16.25} \approx 0.12$$

(d) By Table 5 in Appendix II, $P(z > 0.12) = 0.4522$. For a two-tail test, P value $= 2(0.4522) = 0.9044$.

(e) Since the P value is greater than the level of significance $\alpha = 0.05$, do not reject H_0. At the 5% level of significance, there is insufficient evidence to conclude that the viscosity index distribution has changed with the catalyst.

5. **(a)** Sign test

Sales After	Sales Before	Sign of Difference
610	460	+
150	216	−
790	640	+
288	250	+
715	685	+
465	430	+
280	220	+
640	470	+
500	370	+
118	118	N.D.
265	117	+
365	360	+
93	93	N.D.
217	291	−
280	430	−

(b) $\alpha = 0.01$

H_0: Distributions are the same.

H_1: Distribution after ads is higher.

(c) $x = \dfrac{\text{number of plus signs}}{\text{total number of signs}} = \dfrac{10}{13} \approx 0.77$

$$z = \frac{x - 0.5}{\sqrt{\frac{0.25}{n}}} = \frac{0.77 - 0.5}{\sqrt{\frac{0.25}{13}}} \approx 1.95$$

Use the standard normal distribution.

(d) By Table 5 in Appendix II, the area in the right tail is $P(z > 1.95) = 0.0256$. P value $= 0.0256$.

(e) Since the P value is greater than the level of significance $\alpha = 0.01$, do not reject H_0. At the 1% level of significance, the evidence is insufficient to claim that the distribution is higher after the ads.

7.

Employee	Training Program Rank x	Rank on the Job y	$d = x - y$	d^2
1	8	9	-1	1
2	9	8	1	1
3	7	6	1	1
4	3	7	-4	16
5	6	5	1	1
6	4	1	3	9
7	1	3	-2	4
8	2	4	-2	4
9	5	2	3	9
Sum	45	45	0	46

(a) Spearman rank correlation coefficient test

(b) $\alpha = 0.05$

$H_0: \rho_s = 0$ (There is no monotone relationship)

$H_1: \rho_s > 0$ (There is a monotone-increasing relationship)

(c) $r_s = 1 - \dfrac{6\sum d^2}{n(n^2 - 1)} = 1 - \dfrac{6(46)}{9(81 - 1)} \approx 0.617$

(d) From Table 9 in Appendix II, 0.617 falls between entries 0.600 and 0.700 in row $n = 9$. Use the one-tailed areas to find that the $0.025 < P$ value < 0.050.

(e) Since the P value is less than the level of significance $\alpha = 0.05$, reject H_0.

At the 5% level of significance, we conclude that there is a monotone-increasing relationship between ranks from the training program and the ranks from the job.

9. **(a)** Runs test for randomness

(b) $\alpha = 0.05$

H_0: The symbols are randomly mixed in the sequence.

H_1: The symbols are not randomly mixed in the sequence.

(c) $TTTT|F|TT|FF|TTTT|FFFFFF|TTTTTT;\quad R = 7$

(d) Letting T be the first symbol, $n_1 = 16$ and $n_2 = 9$. From Table 10 in Appendix II, $c_1 = 7$ and $c_2 = 18$.

(e)

$R \leq 7$	$8 \leq R \leq 17$	$R \geq 18$
Reject H_0	Fail to reject H_0	Reject H_0

Since $R = 7$, we reject H_0.

At the 5% level of significance, we conclude that the sequence of answers is not random.

Cumulative Review Problems Chapters 10, 11

1. **(a)** Yes, $\bar{x} \approx 0.61$.

   (b)

   $$P(0) = \frac{e^{-0.61}(0.61)^0}{0!} \approx 0.543$$

   $$P(1) \approx 0.331$$

   $$P(2) \approx 0.101$$

   $$P(3) \approx 0.021$$

   $$P(3 \le x) \approx 1 - 0.543 - 0.331 - 0.101 = 0.025$$

   (c) $\chi^2 = \dfrac{(109-108.6)^2}{108.6} + \dfrac{(65-66.2)^2}{66.2} + \dfrac{(22-20.2)^2}{20.2} + \dfrac{(4-5)^2}{5} = 0.3836$

   (d) H_0: The population has a Poisson distribution with $\lambda = 0.61$
   H_1: The population has a different distribution
   $\alpha = 0.01$
   $d.f. = 3$

   Based on Table 7 in Appendix II, $0.90 < P\text{-value} < 0.95$; do not reject H_0. At the 1% level of significance, the evidence is insufficient to claim that the distribution does not fit the Poisson distribution.

3. **(a)** H_0: $\sigma^2 = 0.3025$
   H_1: $\sigma^2 > 0.3025$
   $\alpha = 0.05$

   Using Minitab, $s^2 \approx 0.3624$. Therefore, $\chi^2 = \dfrac{(n-1)s^2}{\sigma^2} = \dfrac{9 \times 0.3624}{0.3025} = 10.78$ with $d.f. = 9$.

   Using Table 7 in Appendix II, $0.100 < P\text{-Value} < 0.900$; do not reject H_0. Therefore, at the 5% level of significance, there is insufficient evidence to conclude that the standard deviation of *Iris virginica* is greater than 0.3025.

   (b)

   $$\sqrt{\frac{(n-1)s^2}{\chi^2_U}} < \sigma < \sqrt{\frac{(n-1)s^2}{\chi^2_L}}$$

   $$\sqrt{\frac{9 \times 0.3624}{16.92}} < \sigma < \sqrt{\frac{9 \times 0.3624}{3.33}}$$

   $$0.44 < \sigma < 0.99$$

   (c) H_0: $\sigma^2_1 = \sigma^2_2$
   H_1: $\sigma^2_1 > \sigma^2_2$
   $\alpha = 0.01$

   Using Minitab, $s^2_2 \approx 0.1858$, with $d.f._N. = 9$ and $d.f._D. = 7$.

   $$F = \frac{0.3624}{0.1858} \approx 1.95$$

Using Table 8 of Appendix II, P-value > 0.100; do not reject H_0. At the 1% level of significance, there is insufficient evidence to conclude that the variance for *Iris virginica* is greater than the variance for *Iris versicolor*.

5. H_0: The distributions of growths are the same
     H_1: The distributions of growths are different
     $\alpha = 0.01$

$$\mu_R = \frac{11(11+11+1)}{2} = 126.5$$

$$\sigma_R = \sqrt{\frac{11 \times 11 \times (11+11+1)}{12}} \approx 15.23$$

$$z = \frac{135 - 126.5}{15.23} \approx 0.56$$

P-value $= 2P(z > 0.56) \approx 0.5755$; do not reject H_0. At the 1% level of significance, there is insufficient evidence to conclude that the growth distributions are different for the two root stocks.

7.   Using Minitab, the median is 33.45.

H_0: The numbers are random about the median
H_1: The numbers are not random about the median
$\alpha = 0.05$

Moving across the page, *AABBBBAAAABAABBBBA*, so $R = 7$, $n_1 = n_2 = 9$, and using Table 10 of Appendix II, $c_1 = 5$ and $c_2 = 15$. Thus, do not reject H_0. At the 5% level of significance, there is insufficient evidence to conclude that the sunspot activity about the median is not random.